THE COVID 19 PANDEMIC

The Interplay of Medicine, Politics, and Economics in a Global Outbreak

John Uyanne, MD

"Be safe, be smart, be kind."—Tedros Adhanom Ghebreyesus, Director-General, WHO. Press Conference on Covid-19 Pandemic. Geneva.

January 30, 2020.

THE COVID-19 PANDEMIC

PaperBack ISBN: 978-1-957809-34-2

eBook ISBN: 978-1-957809-35-9

Library of Congress Control Number: 2022920736

Published by:

Cornerstone Publishing
A Division of Cornerstone Creativity Group LLC
Info@thecornerstonepublishers.com
www.thecornerstonepublishers.com

Author's Contact

To book the author to speak at your next event or to order bulk copies of this book, please, send email to:

jauyanne@gmail.com

Printed in the United States of America.

DEDICATION

I humbly dedicate this work to the entirety of the front-line workers and responders, without whose herculean efforts, the world might not have seen even a fraction of the success achieved in the battle against the COVID-19 pandemic. By a similar token, I pay deep respect to all those who have lost their lives to the novel coronavirus.

CONTENTS

ACKNOWLEDGMENTS

Words will never suffice in my attempt to express a profound word of gratitude to all those who have, in some form or the other, contributed to the satisfactory outcome of this book, and overall, my eventful and fulfilling career as a practicing physician and teacher. Yet, I consider some individuals and establishments worthy of particular mention.

I acknowledge, with boundless love, the overwhelming support of my wife, Dr. Jettie Uyanne, and my children, Peter and Judith.

I appreciate the sustained support I have received in the course of my career from the following institutions that I am affiliated with: Charles R. Drew University of Medicine and Science, David Geffen School of Medicine at the University of California, Los Angeles, and Martin Luther King Jr Out-Patient Center, Los Angeles.

To all those through whose helpful and impactful paths I have had the immense fortune of crossing over the course of my remarkable career, I express an unreserved gratitude.

AUTHOR'S PREFACE

THE SAGA OF A PANDEMIC

This book presents a comprehensive account of the evolution and spread of the COVID-19 pandemic, also known as the novel coronavirus pandemic. The COVID-19 outbreak, believed to have originated from Wuhan, China, in 2019, spread across the globe in four months. Sadly, the pandemic has taken an unbelievable worldwide toll, leaving behind fatality and morbidity on a global scale. One is either infected or affected as it disrupts virtually everybody's life and lifestyle.

The pandemic has also been accompanied by viral information and misinformation that surpass its capacity for generating fear and anxiety in communities worldwide. As a medical practitioner, I have been accorded the rare opportunity of observing and witnessing the pandemic's effects on people's everyday lives, which is why I can paint a lifelike picture of the destructive capacity and capability of the novel coronavirus.

Accordingly, I have drawn upon my experience and knowledge as a physician, with more than a passing interest in epidemiology, to explain how COVID-19 went from a potentially manageable outbreak to a global pandemic. I critically examined the earliest responses to the pandemic, the history of earlier epidemics, the epidemiological basis of infectious diseases, patterns in the spread of the disease, as well as management and containment of infections at the individual, local, and global levels.

I also explored the pernicious effects of the outbreak on trade and commerce; the social and psychological disruptions that were witnessed on a global scale; the role of the various international organizations and pre-eminent stakeholders in the feverish quest for solutions; and the frantic race for vaccines or a cure. Indeed, these might be considered a verifiable analysis of the COVID-19 pandemic and its global response.

I have also beamed a ray of hope into the future, while highlighting clear steps governments knew they could have taken to prevent or prepare for the present pandemic. I did not shy away from making a bold and optimistic case for a pandemic that may finally inspire the world to take viruses more seriously. Of course, fighting this pandemic and preventing the next one will take political action from governments, the scientific community, and individuals; but it is possible.

In all, I have tried to make this a one-stop resource that ought to be an indispensable tool for the average reader, a holistic storehouse of knowledge for researchers in the fields of public health, medical sociology, political economy, public policy and governance, medicine, and paramedicine, and an invaluable resource for policymakers and civil society organizations. The book educates the reader about the basic principles of infection, epidemics, and pandemics, including information about the novel coronavirus, the classic epidemiological triad, infection chains, and the progression from an outbreak to a pandemic. In addition, it delves into the history of the recent pandemics that have afflicted humanity, reminding the world that it abysmally failed to learn from history.

Moreover, the book details information about the deadly Spanish flu of the early twentieth century, as well as SARS, MERS, Swine Flu, and Ebola, while describing how the COVID-19 pandemic originated in China and raged like a forest fire across the globe, leaving unimaginable misery afterward. Also, it gives an account of how the pandemic's epicenter quickly shifted from China to Europe and the United States; how these countries tried to contain it, and the spread of the pandemic in gravely affected areas like India, Brazil, Turkey, Russia, and Iran.

The concise description of the features of COVID-19 includes the case definition the WHO provided and clinical features of and treatments for COVID-19. Perhaps the most heartbreaking sections of the book are those that deal with the impact of the COVID-19 pandemic on humanity. The toll on mental health is indescribable, while the gross disruptions to social life, dislocations in community wellbeing, and severe devastation to the economy leave one gasping in sheer disbelief. Also, the reader will be familiar with the methods for containing the pandemic—social distancing, self-isolation, hygienic measures, quarantine, lockdown of cities, the suspension of international travel, and border closure.

The closing chapters explore global issues that have emerged from the pandemic, which include the frantic race to develop effective vaccines and the preparedness and strengthening of public health infrastructure. They also include information related to new variants of the coronavirus that continue to emerge, even as the manuscript went to press. Generally, the book is presented in a few chapters, using plain and comprehensible language. I also provided references to aid further and enlightened reading and a few illustrations to enrich the reading experience.

In conclusion, this work relies heavily on information provided by the WHO, the United States Centers for Disease Control and Prevention (CDC), highly respected medical journals, and several credible international media organizations. It is my hope that my modest effort has brought a comprehensive, informative, and accessible knowledge of COVID-19, which requires reading for contemporary times and beyond. It is too early to make conclusive and dogmatic opinions on where the COVID-19 pandemic is taking us, but it is certainly past the time to continue dwelling on what went wrong.

The time to learn about how we can do better has come, and it is now.

John Uyanne, MD
Los Angeles, California
October, 2022.

THE OVERVIEW

Chronicling a Pandemic

The world is, undoubtedly, in a global battle against novel strains of the coronavirus. The first and new strain of coronavirus, named SARS-CoV-2 by the *International Committee on Taxonomy of Viruses,* was first reported in Wuhan, in Hubei province, China, in December 2019; and it has spread rapidly across almost all regions of the world. The WHO named the disease COVID-19, a shortened version of Coronavirus Disease 2019, which SARS-CoV-2 infection causes. The virus has genetic kinship with the SARS-CoV virus that was responsible for the Severe Acute Respiratory Syndrome (SARS) epidemic of 2002–2003.

COVID-19 has resulted in a harvest of deaths and severe illnesses and also affected virtually every aspect of human life. The massive disruption in daily lives is such that entire cities and countries have come under bizarre lockdown. In addition, many international sporting events, social events, marriages, and other ceremonies have had to be postponed or canceled, resulting in a global economic crisis of incredible magnitude.

Every industry has been affected on the global turf; stock markets have crashed like a pack of cards. The world's hospitality industry, air transportation, and the entire travel and tourism sector have also been crippled. Naturally, it has led to sudden widespread unemployment across all sectors of the economy.

That the world was caught unawares is a mere understatement. Patent lessons that ought to have been learned from the past epidemics were either taken with extreme levity or not heeded. Obvious warning signs were also not given adequate and deserving attention. Tragically, healthcare workers, essential service providers, and frontline first responders suffered the first major casualties, trying their heroic best to fight the deadly virus with little protective gear while they succumb to its ravaging effect at an alarming rate.

Although past epidemics were associated with high-density human neighborhoods, lack of adequate healthcare facilities, poor environmental hygiene, and poor sanitation facilities, people did not know them to spread at such an alarming global rate. However, travels by ship and air were involved in the progression of a local outbreak to the global spread, termed a pandemic.

For instance, in 1968, influenza became the first pandemic to spread through air travel. It was accompanied by an epidemic of acute hemorrhagic conjunctivitis that spread between international airports. That was the dawn of a new and modern era of epidemics and rapid transnational infection spread. We now know that human migrations, rapid population rise, poor sanitation facilities, poverty, and the negative impact of human activity on the ecosystem have enhanced the likelihood of infectious diseases and the associated evolution of pandemics in recent history. In a nutshell, mankind has probably turned out to be it's worst enemy.

The WHO reports that, by May 15, 2020, there were 4,338,658 confirmed cases of COVID-19, while the death toll had reached 297,119 people globally. Yet, there are discrepancies in the data from reputable institutions like Johns Hopkins University and the WHO. We can identify many reasons for these discrepancies—data figures are reflected in how a case of COVID-19 is defined and in the methodology for collecting and collating data.

The unavailability of test kits at the nascent stage of the pandemic marred efforts at tabulating data. As a result, the world was already suffering casualties even before we could arrive at the right diagnosis. However, it remains a fact that the actual numbers will probably never be fully known. Yet, we are constrained to accept that data do not always tell the entire story of the misery, death, and despair wreaked on humanity by a deadly global pandemic.

The WHO declared the coronavirus outbreak *a public health emergency of international concern* on January 30, 2020, and *a pandemic* on March 11, 2020. This declaration and the worldwide harvest of death, especially in developed countries of the West, resulted in relentless fear and panic. To worsen the situation, there were no identifiable medicines and vaccines to treat and prevent a novel coronavirus infection. Moreover, the world had rarely been caught so unawares, as the global medical community possessed little or no knowledge of the novel coronavirus at the beginning of the attack.

The role of China, where the outbreak started, and the WHO in mitigating the COVID-19 pandemic became a source of intense controversy. China continued to face a global backlash over what was perceived as the initial mismanagement of the outbreak; Australia insisted on an inquiry into the origin of the coronavirus infection; Donald Trump, the 45th president of the United States, also blamed China for the contagion. Simultaneously, the WHO faced mounting pressures from critics who accused it of succumbing to China's influence to compromise and hinder effective response to the coronavirus pandemic.

President Donald Trump ceased funding to the WHO over its handling of the coronavirus pandemic, but many people expressed the rational view that it was not the time to play the blame game. Rather, it was the time for dedicated and concentrated efforts to contain the damage and halt the spread of the pandemic.

This overview is intended to provide a sequential layout of the evolution of COVID-19 presented in this book. I sieved fact from fiction and furnished an accurate picture of a unique human crisis while presenting it in chapters that offer virtually all information on the COVID-19 pandemic. It covers the basic principles underlying infectious disease, how an infective entity progresses into a pandemic, the fundamental information on epidemics, pandemics, pathogens, immunology, infection chains, and the transmission of infection while providing insight into case fatality rate, incubation periods, and reproductive number (R_0) to help the readers understand the pandemic and its dynamics.

The book also delves into the various pandemics that have attacked our world in the past, including the Spanish flu, SARS, MERS, swine flu, Ebola, and Zika. It also offers brief treatises on how these pandemics started and impacted humanity, enabling readers to compare past pandemics with COVID-19.

Moreover, the book describes the origin and spread of the COVID-19 pandemic and the global response to it. We know that the entire epidemiological saga of COVID-19 originated at the Huanan Seafood Wholesale Market in Wuhan, China, from where it spread rapidly among the residents of that locality. There was a massive propagation in the spread, with mounting casualty figures, resulting in a massive mobilization of healthcare resources to contain the outbreak.

The coronavirus had spread to various countries, within a twinkling of an eye, with the WHO declaring it a pandemic. In this book, I gave a timeline of the evolution of the epidemic, how it transformed from an epidemic to a pandemic in a short time that saw its epicenter shifting rapidly from China to the sophisticated economies of Europe and the United States, and the evolution of the disease and the number of cases and fatalities in countries like China (where it started), Spain, Italy, France, the United Kingdom, Germany, Australia, Canada, and the United States, which became its major epicenter within months.

The chronological journey takes us into the spread of the pandemic in Brazil, Turkey, Russia, India, Iran, and the rest of the world. Russia, Brazil, and Turkey quickly turned out among the worst-affected nations in the world. India became severely affected that it left its government almost helpless with resignation, with its teeming underserved populations.

Furthermore, I described the COVID-19's signs and symptoms in explicit detail, focusing on how COVID-19 started with fever and cough before it rapidly evolved into pneumonia, respiratory failure, and death in a very short time. This book will educate the readers on the common and uncommon symptoms of the disease, basic personal intervention, infection prevention protocol, laboratory assessment to make an informed diagnosis, and vital health tips.

Also, I explained the various governments' and financial institutions' efforts to buoy national economies. Finally, I described the mental distress and socioeconomic dislocations that COVID-19 caused without leaving out the gross and overall impact of the pandemic on trade and commerce, international sporting events, the tremendous psychological effects on people, and the disruptions in their daily lives.

Containment of the coronavirus pandemic has been a monumental challenge. Initially, there was no effective medicine or vaccine. Therefore, infection prevention became the most important aspect of containment in the devastating pandemic. This book discusses the steps for preventing infection and containing the pandemic from personal, national, and global perspectives. It also emphasizes quarantine and locking down cities, annulment of visas, cancelations of air, train, and ship travels, and other drastic strategies to control the movement of people, which played a significant role in the containment of the pandemic. The game's name became *flatten the epidemic curve* so that health facilities would not be overwhelmed.

Many global challenges emerged due to the COVID-19 pandemic. For example, efforts at vaccine development and strengthening the public health system became vital issues. Significantly, those critical global challenges will be important in the current pandemic and the future. The final chapters provide the latest information on the COVID-19 pandemic. The pandemic should sensitize us on the possibility of the emergence and regeneration of infectious agents and the need for constant vigilance. Nothing can take the place of effective countermeasures if humanity should be spared the sort of agony it has witnessed with the COVID-19 pandemic.

CHAPTER ONE

PANDEMIC: DEFINITION AND FUNDAMENTAL PRINCIPLES

The onset of the novel coronavirus infection worldwide has brought terminologies such as *epidemic*, *endemic*, *pandemic*, *incubation period*, *case fatality rate*, etc., that many people are unfamiliar with. While several people might have come across these terms, perhaps through cursory reading, they may neither have a complete understanding of these words nor have any clue about their involvement with a deadly ailment like COVID-19. In this chapter, I defined the trending terminology surrounding the evolution and spread of the novel coronavirus. However, it is vital to know the difference between these words: *epidemic* and *pandemic*.

Outbreak

When an infectious disease remains localized to a small geographic area, it is called an outbreak; it possesses the same significance as an epidemic.

Epidemic

An epidemic is an unexpected increase, often spontaneous, in the number of cases of an infectious disease in the population of a given area over a specified period.

Pandemic

A pandemic is an epidemic that has spread over several countries or continents, usually affecting many people. For example, the coronavirus infection was an outbreak when it was localized to the city of Wuhan, China. It became an epidemic when it spread rapidly to involve many people in several cities in China. However, it became a pandemic when it crossed China's borders and spread to several countries and continents, affecting a huge population.

Sporadic, Endemic, and Hyperendemic

A disease that occurs irregularly is termed *sporadic*. Also, disease or any infectious agent that remains prevalent in a population within a particular geographical area is said to be *endemic* to that area. However, it becomes *hyperendemic* if the disease becomes high and persistent within that area.

Incubation period

There is usually an interval between exposure of a susceptible host, person, or animal, to an infectious agent, such as a virus, bacteria, or protozoa and the appearance of the first signs or clinical symptoms of the disease in that host. This period is referred to as the incubation period of a disease. According to the WHO, the incubation period for COVID-19 ranges from one to fourteen days, with the most common average of around five days. This means that after a healthy person comes into contact with someone with a coronavirus infection, he or she is likely to develop an infection within fourteen days of contact.

Case Fatality Rate

The case fatality rate is an important, empirical form of assessment of the progression of an epidemic or pandemic. It is the proportion of deaths concerning the number of reported cases of a disease. In other words, it is the number of people that a disease infected within a specified time. The case fatality rate of the coronavirus pandemic has regional variations, and it varies by age, sex, and the presence or absence of comorbidities such as diabetes, hypertension, respiratory diseases, etc.

The overall fatality rate of people with confirmed COVID-19 in the Italian population, based on data up to March 17, 2020, was 7.2% (1625 deaths/22,512 cases). The overall case fatality rate in Italy (7.2%) was significantly higher than that in China (2.3%), while the case fatality rates in Italy and China appear similar for age groups 0 to 69. Still, rates are higher in Italy among individuals aged 70 years or older, particularly those aged 80 years and above.

Yet, exact case fatality rates are difficult to get because not all countries possess adequate testing capabilities to detect the coronavirus. Additionally, the pandemic is evolving, and the ultimate result may differ from what is currently obtained.

Basic Reproduction Number (R_0)

Another key concept used in this era of the coronavirus pandemic is the basic reproduction number, R_0, which was previously called the *basic reproductive ratio* or *rate*. In terms of the reproduction number, there are three possible situations:

$R_0 = 1$ indicates that the number of cases is stable and the disease is endemic.

$R_0 > 1$ indicates that the number of cases is increasing, and it will ultimately turn into an epidemic.

$R_0 < 1$ indicates that the number of cases is decreasing, and the epidemic will be under control soon.

A few months after the onset of the pandemic, the R_0 for COVID-19 in China was 2–2.5, which means that each patient transmitted the infection to an additional 2.2 individuals. In Italy, the basic reproduction number, at that time, ranged from 2·76 to 3·25. These data indicate that coronavirus infection is highly contagious, and the epidemic may take some time to stabilize.

Immunity

Immunity refers to the development of resistance against an infectious agent. When vaccination or natural infection develops immunity, it is known as *active immunity*; but when it is transferred from mother to fetus or by administering specific protective antibodies, it is known as *passive immunity*.

Natural immunity implies the innate immunity of human beings to specific disease-causing agents. *Herd immunity* refers to the percentage of individuals in a population who have developed immunity against a specific disease-causing agent. Sadly, the British government did not act to prevent large gatherings of people at the onset of the pandemic. Instead, that strategy was to generate herd immunity to fight the coronavirus epidemic.

However, immediately after the British health authorities understood the high fatality rate of the infection, the government shelved the strategy. Since SARS-CoV-2, the coronavirus causing COVID-19, is a novel pathogen, there is little immunity in large areas of the population against this disease, and almost everyone is susceptible to the coronavirus infection.

COVID-19

COVID-19 is an infectious disease. Therefore, a basic understanding of infectious diseases is appropriate at this point.

Different disease-causing agents that enter a human's or animal's body and affect one or more vital body parts, such as the lungs, intestines, or urinary tract, causes infectious diseases.

These organisms, which are everywhere, can enter the body through the nose, mouth, or skin; they can multiply afterward, which may or may not lead to an illness.

Epidemiological Triad

A traditional model used to understand the progression of infectious diseases, such as COVID-19, in the human body is called an *epidemiological triad* or *triangle* (See figure below). This triad consists of a disease-causing agent, a susceptible host (a human being), and an environment that connects the agent and the host. A disease occurs when the agent interacts with the susceptible host in an environment that supports the transmission of the agent into the host's body.

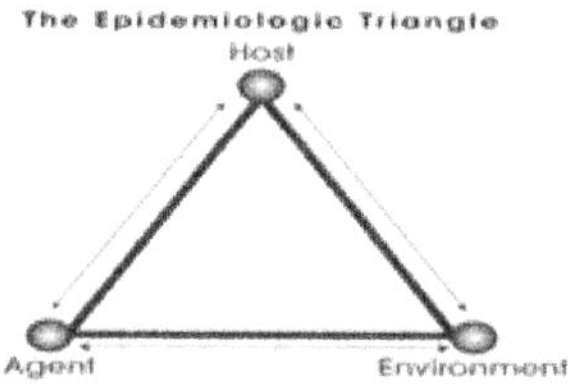

Agents

Agents, also called pathogens, are microorganisms that cause diseases. They include bacteria, viruses, fungi, and protozoa. The development of a disease depends on several factors related to these agents; for instance, their pathogenicity, which is their ability to cause disease, and infecting dose or quantity. The mere presence of microorganisms in the environment is insufficient to cause disease. Instead, the greater the number of pathogens in the body, the more severe the disease.

Bacteria

Bacteria are disease-causing microorganisms. they causes cholera, streptococcal sore throat, and typhoid. In addition, bacteria can reproduce by themselves, thereby increasing the severity of the disease they cause when they enter the human body.

Fungi

Fungi are another type of microscopic pathogens that are involved in many skin and lung infections. They are often compared with plants since they comprise many cells, unlike bacteria or viruses. However, because they cannot produce food by themselves, they depend on human beings and plants for survival; and they cause infection when the host has a low immune status.

Protozoa

Protozoa are single-celled organisms found worldwide. Most protozoa species are free-living; they have complex internal structures and carry out complex metabolic activities. Some species are considered commensals, which means they are not harmful, while others are pathogens and usually cause diseases. The infections they cause range from asymptomatic to life-threatening conditions e.g. malaria.

Viruses

Viruses are microscopic organisms that are smaller than bacteria. They are responsible for diseases like measles, chicken pox, and the common cold. An important difference between a virus and a bacterium is that a virus cannot replicate itself, although it has its genetic material, while a bacterium can replicate itself.

Once a virus enters a human body, it attaches itself to the cell surface with a receptor and gains entry to the cell. It then uses the cell's machinery to multiply itself severally.

Coronaviruses are a special type of spherical virus made up of enveloped particles containing single-stranded RNA. The envelope bears club-shaped glycoprotein projections. They possess a crown-like or halo-like appearance of the envelope glycoproteins under an electron microscope.

Coronaviruses are responsible for the common cold. They are transmitted from one host to another via airborne droplets. The virus reaches the host's nose and invades his respiratory tract, where it starts multiplying locally in the cells, leading to cell inflammation and damage. However, these viruses usually do not cause much damage to the lungs.

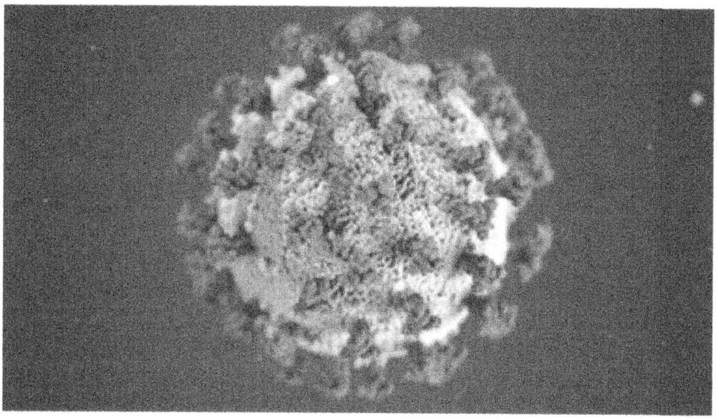

The infection leads to the production of antibodies, which ultimately eliminate the virus from the body; and the infection subsides. The presence of the antibody may last for a year or two to provide lasting immunity for future infections. However, there is greater infectivity during winter, which may appear as a local epidemic. Diagnosis may be difficult and hard to differentiate from other common cold causes, but treatment is symptomatic, as there is no specific therapeutic remedy.

The coronavirus has gained notoriety in this century because it has been responsible for an epidemic of massive proportions. In addition, these viruses can cross specie barriers to cause severe diseases such as SARS and MERS due to some reasons. However, the only reasonable theory is that these viruses probably originated from bats and migrated into other mammalian hosts, perhaps the Himalayan palm civet for SARS-CoV and the dromedary camel for MERS-CoV, before jumping to humans.

SARS-CoV-2 is a new strain of coronavirus that seems very contagious and has quickly spread globally. It belongs to the betaCoVs category with a diameter of approximately 60–140 nanometers, and it is susceptible to heat, ultraviolet rays, ethanol, 75% ether, and chlorine-containing disinfectants. The isolates from the Wuhan strain share 89% nucleotide identity with bat SARS-like-CoVZXC21 and 82% with human SARS-CoV-12; hence, it was named SARS-CoV-2.

The SARS-CoV-2 spike protein binds with the human cell surface of the angiotensin-converting enzyme-2 (ACE2) receptor to gain entry into the cell. ACE2 is expressed by the epithelial cells of the lungs, intestine, kidney, and blood vessels. It causes damage to the cells, leading to inflammation and extensive tissue damage after it enters the cell. It causes mild symptoms, and the patient may be asymptomatic in most cases; sometimes, it causes severe lung damage, leading to death.

Host

In the epidemiological triad described above, the host is a human being or an animal that harbors disease-causing pathogens. The host may become ill or remain asymptomatic, carrying the agent silently and spreading it in the community. Yet, people are not equally susceptible to the disease because several risk factors intrinsic to the host determine the susceptibility or otherwise to the disease. These significant risk factors are age, sex, race, immune status, smoking and drinking habits, drug addiction, occupation, nutritional status, genetic composition, comorbidity, and psychological profile.

Studies indicate that coronavirus infection is more likely to affect older male patients. The median age of patients with COVID-19 ranges from 49 to 59 years. Infection in children is reported to be less common than in adults. At least, cases have generally been in family clusters or children with a history of close contact with an infected adult at the early stage of the pandemic.

Most COVID-19 patients have been found to have underlying comorbidities like diabetes, hypertension, respiratory disease, or kidney dysfunction. Clinical and epidemiological data from the Chinese CDC (Centers for Disease Control and Prevention), regarding 72,314 case records reported in the *Journal of the American Medical Association,* has revealed that the fatal cases were primarily elderly patients, in particular those over 80 years (about 15%) and those aged 70 to 79 years (8.0%). About half (49.0%) of the severe patients with pre-existing comorbidities, such as cardiovascular disease, diabetes, chronic respiratory disease, hypertension, and cancer, died.

Similarly, data from the United States revealed that between February 12 and March 16, 2020, 4,226 COVID-19 cases were reported in the United States. Among them, 31% of cases seem unclear, 45% of hospitalizations, 53% of Intensive Care Unit (ICU) admissions, and 80% of deaths occurred among adults aged 65 and above, with the highest percentage of severe outcomes among adults aged 85 and above.

These data are similar to the data obtained from the Chinese CDC. Yet, it does not mean that young people are not suffering or are not being hospitalized. For example, the same report of the US CDC revealed that among the hospitalized patients, 20% were aged 20–44; among all the patients needing ICU care, 12% were in the age group 20–44.

French health ministry official Jérome Salomon said, in an article published in Washington Post on March 19, 2020, that half of the 300 to 400 coronavirus patients treated in intensive care units in Paris were younger than 65; and according to numbers presented at a seminar of intensive care specialists, half the ICU patients in the Netherlands were younger than 50. The Italians reported that most patients are from the older age group (age more than 60 years), and maximum death has also occurred in the older age group (individuals aged 70 years and 80 years or older). Italy has a high proportion of older patients with confirmed COVID-19 infection. The fact that the elderly in Italy account for a significant number of the population, with approximately 23% of the Italian population aged 65 years or older, may explain the high mortality rate in this age group.

Environment

The environment comprises all the external factors that affect the host, the agent, and their interactions. For example, the environment includes physical factors such as humidity, temperature, and geology; socioeconomic indices such as poverty, pollution, sanitation, crowding, availability of health services, and psychological makeup and biological factors such as insects that transmit the agent to cause disease.

That is why the doctor must consider the patient in the total population to which that patient belongs to properly understand the dynamics of any disease in a particular environment. Infectious diseases often do not occur in isolation but a group of people exposed to the agent from a single source. Early cases in

Wuhan were believed to have acquired infection from an animal source since many of them had reported visiting or working in the Huanan Wholesale Seafood Market.

Yet, travel history is another environmental factor that demands vital consideration in the current pandemic. People with a travel history to China, Italy, France, Iran, and other coronavirus-affected countries were screened; many were quarantined in India and many other countries to control the spread of infection through early detection and isolation. Several factors like overcrowding, environmental changes, lack of adequate healthcare facilities, and air travel have led to the development of a human-driven ecosystem that provides favorable conditions for the emergence of rapidly mutating RNA viruses, which causes havoc and threats to humans.

Prevention

We need to understand the complex interaction between the agent, host, and environmental factors and devise strategies to halt the interaction between them, to prevent or control infectious diseases. For instance, social distancing and social isolation have been effectively used as strategies to control the spread of infection during the coronavirus pandemic. In addition, these strategies broke the chain of transmission of coronavirus from one person to another.

From Infection to Pandemic

COVID-19 became an epidemic and a pandemic because the coronavirus and its susceptible hosts were present in adequate numbers in an environment, and the coronavirus was effectively and rapidly transmitted to the susceptible hosts.

An epidemic may result from the following:

1. Entry of a new pathogen into an environment where it was previously absent.

2. A recent and sudden increase in the pathogen's level or virulence.

3. An enhanced transmission mode, such that more susceptible people are exposed.

4. A change in the susceptibility of the host to pathos.

5. The presence of factors that increase the host's exposure or involve introduction through new portals of entry.

Once an infectious agent, for example, influenza, establishes its presence in an environment, it continues to infect several people; and the disease spreads from one person or location to another and across the borders of the country into the neighboring country until it becomes a pandemic. However, it can be expected to eventually settle into a cycle in which susceptible people are infected and become immune to further attacks.

The virus moves on as the number of susceptible people reduces and returns when the next generation of susceptible populations, including those born after the last epidemic, appear in the environment. The other reasons for a cyclical trend of an epidemic are the emergence of a new group of susceptible populations due to the decline in protective immunity and the removal of immune individuals due to death.

Transmission of Infection

Transmission of infection is significant in the progression of a disease. Therefore, it is important to have a thorough familiarity with the mode of transmission of an infectious agent to understand the evolution of a disease and its transformation into an epidemic or pandemic, which helps to determine the strategies to contain the spread.

A chain of infection is established when an infectious agent comes out from its reservoir or host through a *portal of exit*, gets carried by some mode of transmission, and enters the body of a susceptible host through an appropriate *portal of entry*. Establishing a chain of infection is necessary for an infection to occur. However, the maintenance of this chain and its quick spread lead to an epidemic.

Reservoir

A pathogen's reservoir is the habitat in which it finds suitable conditions that permit it to reside, grow, and multiply for a long time. The reservoir may be living or non-living and may include humans, animals, or the environment. For example, common infectious diseases like measles and mumps have human reservoirs.

A carrier is a person who harbors the pathogen and can transmit the pathogen to others but is not ill from the disease. Such asymptomatic healthy carriers are important transmitters of pathogens to other susceptible hosts. The coronavirus pandemic has spread via these types of healthy carriers. Many young people become infected but do not show any signs of illness. Naturally, they are neither suspected to be carriers nor tested to see whether they are carriers. Instead, they move around, infecting others. The WHO report highlights that out of some 3,664 cases reported outside China, 92 were detected while apparently asymptomatic.

Similarly, many pathogens reside in animals and occasionally jump from animals to humans. HIV, Ebola, and SARS evolved from animals to cause widespread disease. The ecological reservoir of SARS-CoV-2 is bats, from which the disease spreads to human beings.

Portal of entry

The portal of entry is the entry site of a pathogen into a susceptible host's body. The respiratory tract is the portal of entry of other coronavirus-like respiratory pathogens.

Portal of Exit

A Portal of exit is the path a pathogen takes to exit from a host's body. For example, Mycobacterium tuberculosis exits via the respiratory tract, while Vibrio cholerae exits through feces. The portal of exit for SARS-CoV-2 is mainly the nose and mouth. However, further research is required to identify all the possible portals of exit.

Mode of Transmission

1. Direct contact and droplet spread

These are two major modes of transmission of coronavirus infection. Droplet spread implies the spray of pathogens through the large, short-range aerosols produced via sneezing, coughing, or talking. Droplet spread transmits the pathogen by direct spray over a few feet before the droplets fall to the ground.

We know that close contact between two people—an infected and a healthy person—is unnegotiable for SARS-CoV-2 to spread. Here, physical contact between the two individuals through interpersonal actions like touching, kissing, shaking hands, or hugging can transmit the pathogen. The virus can also spread directly from one person to another when a COVID-19 patient coughs or exhales, producing droplets that arrive at another person's nose, mouth, or eyes.

Alternatively, as the droplets are too heavy to be airborne, they land on objects and surfaces in the vicinity of the person. Other people then become infected with COVID-19 by touching these contaminated objects or surfaces and touching their eyes, nose, or mouth afterward. That is why COVID-19 spreads rapidly among members of the same family, healthcare professionals treating such patients, and other people who come close to an infected person.

2. Airborne transmission

Airborne transmission occurs when dust particles or droplet nuclei suspended in the air carry pathogens. Research shows that the SARS-CoV-2 virus can live in the air for up to three hours. Therefore, it can get into your lungs if an infected person breathes out and you breathe in that air. However, experts are divided on how often the virus spreads through the airborne route and how much it has contributed to the pandemic. Given the many similarities between the two SARS viruses and the evidence of virus transport, it is highly likely that the SARS-CoV-2 virus also spreads by air.

3. Vehicle-borne transmission

Food, water, blood and fomites, and inanimate objects like handkerchiefs, masks, towels, and surgical instruments, can transmit pathogens. The US CDC reports that the novel coronavirus has not been detected in drinking water. Also, there have been no reports of fecal-oral transmission of COVID-19, although it has been detected in the feces of many COVID-19 patients. However, healthcare workers have detected and reported viable SARS-CoV-2 in COVID-19 patients' stool and virus RNA in sewage, raising the likely specter of fecal-oral transmission.

There is significant evidence that COVID-19 is transmitted to close contacts by fomites. Therefore, towels, handkerchiefs, and masks are likely sources of infection to family members and healthcare workers. So naturally, properly disposing of used masks, gloves, and gowns is crucial in preventing the spread of the virus.

4. Vector-borne transmission

Vectors like flies, fleas, ticks, and mosquitoes carry pathogens and transmit diseases. An example is the transmission of malaria by mosquitoes. However, there is no evidence of vector-borne transmission of COVID-19 to date.

Susceptible Host

A susceptible host is a final link in the chain of infection. Susceptibility depends on genetic, nutritional, immunological, or quite unspecific factors. Unfortunately, there is hardly any immunity against SARS-CoV-2 in the general population since it is a new strain of coronavirus. As a result, the infection spreads rapidly and widely, quickly turning into a pandemic.

The Pandemic Phases

The WHO defined the pandemic phases in 1999 and revised them in 2005 in the context of an influenza pandemic to help countries in preparation and response planning. These phases help assess the spread of a pandemic and implement mitigation measures. The six phases are as follows:

Phase 1 – No viruses circulating among animals have been reported to cause human infections.

Phase 2 – An animal influenza virus circulating among domesticated or wild animals is known to have caused a human infection. Hence, it is considered a potential pandemic threat.

Phase 3 – An animal or human-animal influenza virus has caused sporadic cases or small clusters of disease in people. However, it has not resulted in human-to-human transmission sufficient to sustain community-level outbreaks.

Phase 4 – Human-to-human transmission of an animal or human-animal influenza virus that can cause community-level outbreaks has been verified.

Phase 5 – Human-to-human spread of the virus into at least two countries in one WHO region.

Phase 6 – Community-level outbreaks in at least one other country in a different WHO region, in addition to the criteria in Phase 5.

For the COVID-19 pandemic, the mutated coronavirus acts as a deadly agent that attacks susceptible hosts with no immunity to the disease. The virus spreads rapidly from one person to another by droplet spread and direct contact. This chain of transmission has resulted in an outbreak that quickly progressed to an epidemic, especially because there were no vaccines or medicines to alleviate the disease.

Additionally, global mobility has carried the virus from its origin to the nooks and crannies of the world, turning a serious epidemic into a deadly pandemic. Humanity is no stranger to pandemics. A century ago, the Spanish flu pandemic spread through almost every country worldwide, while the SARS pandemic spread worldwide in 2002-2003. The similarity is that these pandemics started in some parts of the world, and due to the extreme infectivity of the pathogen, they spread rapidly from one country to another with harmful consequences.

CHAPTER TWO

THE MEMORY LANE OF THE VIRUS WORLD

COVID-19 is not the world's first experience with a raging pandemic. Infact, mankind had borne the brunt of infectious diseases for a millennia. At some point, those scourges were the leading cause of human attrition across the ancient world and the world in recent history.

Mankind's fate changed positively with the discovery of penicillin in 1928 by Sir Alexander Fleming. Penicillin emerged as the new and potent weapon for treating infectious diseases. When doctors started using the drug in the 1940s, it put up a formidable battle against various infectious diseases and became one of the most sought-after medicines in that era.

Infectious diseases remain a major cause of morbidity and fatality in our modern world, despite decades of research and development and a sustained effort that has increased the capacity of doctors to prevent and manage them.

With the advent of advanced antibiotics, several doctors believed that infectious diseases would soon become relics of the past, but they were mistaken. While it is true that newer medicines were developed that were found to be effective against many pathogens, the microbes soon gradually developed resistance against such medicines.

Antibiotic resistance developed rapidly to match the pace of the development of new antibiotics. As a result, many diseases that experts once believed could be eliminated re-emerged with huge force. Tuberculosis is an excellent example. In addition, the frequent interaction between humans and animals has recently added a new dimension to developing infectious diseases.

Pathogens that were the exclusive preserve of animals have gained a foothold in humans. Environmental changes and the proximity of humans to animals are responsible for this alarming trend. However, an assessment of the deadly pandemics that have wreaked unspeakable havoc on humanity in the past shows the destructive potential of these new diseases, leaving medical science clueless about prevention and cure.

Such episodes also prove that we will continue to witness the emergence of similar deadly diseases in the future, and the medical community will continue to find it difficult to deal with such disasters whenever and wherever they appear. No amount of preparation would be enough to control such outbreaks. Notwithstanding, we can make a renewed effort to learn from these episodes and do our best to develop emergency strategies to help minimize these infectious diseases' impact.

Spanish Flu

The Spanish flu started in 1918. The *Bulletin of the History of Medicine* documents that the pandemic resulted in about 50 million deaths worldwide. Influenza, commonly known as the flu, is an acute respiratory illness that influenza viruses A and B cause. New strains of influenza, for which people have no immunity, occasionally appear, causing pandemics that affect many people within a short time. There have been thirty-one known influenza pandemics since the first well-documented pandemic of 1580, including three pandemics during the 20th century (1918, 1957, and 1969).

According to the WHO, Spanish flu, the pandemic of 1918-1919, was particularly virulent and killed an estimated 40 million people worldwide. The Spanish flu did not originate in Spain, despite its nominal identity; it likely originated in the United States. It was given the name because Spanish newspapers were the first to report the outbreak. Spain was a neutral nation during World War I, so its journalists did not face strict censorship, unlike those working in countries involved in the war.

The microbe circled the globe in four months, claiming millions of lives. The United States lost 675,000 people to the Spanish flu of 1918, garnering more casualties than it did in World War I, World War II, the Korean War, and the Vietnam War.

American forts had experienced a remarkable expansion at the imminence of war. At Fort Riley, Kansas, a new training facility Camp Funston built to house some of the 50,000 men who would later be drafted into the army, was devastated by the Spanish flu. In early March of 1918, a soldier reported to the camp infirmary in the throes of a fever. Within a few hours, more than a hundred other soldiers came down with a similar ailment, and more fell ill over the following weeks.

With the disease subsiding in the United States, it crept into the battlefields of Europe. The disease was named with various colorful names. Americans fell ill with a *three-day fever* or *purple death while t*he French caught *purulent bronchitis*. The Italians had a *sand fly fever*, while German hospitals were filled with victims of *Blitzkatarrh* or *Flanders fever.*

Yet, whatever name it received in whatever part of the world, its clinical features bore close similarities in each location. It started with a fever, sore throat, and chills and rapidly involved the lungs. Fluid poured into the lungs because of severe inflammation, and patients died from respiratory failure. Patients drowned in their lung secretions because the lungs were swamped with fluid. The flu struck America with high fatality figures. The United States was awash with stories of people waking sick and dying on their way to work, overcrowded hospitals, and mass graves.

Gradually, the flu reached the shores of India and China. The pandemic left almost no part of the world untouched. It is estimated that in the United Kingdom, 228,000 people died; in Japan, around 400,000 people died. The South Pacific Island of Western Samoa, now The Independent State of Samoa, lost about one-fifth of its entire population. Researchers estimated that in India, fatalities totaled between 12 million and 17 million. Although exact data are absent, global mortality figures are estimated to have been between 10% and 20% of infected people.

Sadly, that pandemic counted as the greatest human massacre of the 20th century. Yet, it has been buried in the sands of history since hardly anyone remembers it today as a pandemic. Only a few people remember it as a private tragedy in their family history.

However, the search for this deadly disease was sustained. Researchers extracted lung tissue from the body of a woman who had died of the Spanish flu, buried deep under layers of ice in Alaska. They isolated the influenza virus in 2005, cultivated it in cell culture, and studied its genome. They eventually found the virus to be of avian origin and must have jumped the species barrier to cause the deadly Spanish flu.

They studied this virus in mice and found that its multiplication rate was much higher than other flu viruses. Moreover, it attacked the lungs specifically, where its concentration was 39,000 times greater than that of other flu viruses. Dr. Tumpey T.M. and his colleagues later wrote that *"the constellation of all eight genes together make an exceptionally virulent virus."*

SARS

SARS is an acronym for *Severe Acute Respiratory Syndrome.* It first appeared in November 2002 in the Guangdong Province of Southern China as a viral respiratory illness that a new strain of coronavirus caused. With the number of recorded cases increasing exponentially in a very short timeframe, it soon became an epidemic.

From 2002–2003, it involved four continents and 87 countries, resulting in about 744 deaths. However, according to the WHO, SARS affected 26 countries and resulted in more than 8,000 cases in 2003. The difference between the data that WHO and other authorities provided is due to different methods of case diagnosis, data collection, and reporting. The SARS coronavirus (SARS-CoV), an animal virus from an as-yet-unascertained animal reservoir, possibly the bat, that spreads to other animals, most probably the civet cat, was identified in 2003.

SARS is thought to be transmitted most readily by respiratory droplets emitted when an infected person coughs or sneezes, bearing many similarities to COVID-19. However, the virus can also spread when a person touches a surface or object contaminated with infectious droplets and then touches the mouth, nose, or eyes. In addition, the SARS virus might spread more broadly through the air or other unknown means. Due to inadequate infection control precautions, most human-to-human transmission cases occur in healthcare facilities.

The common symptoms of SARS were high fever, general malaise, myalgia (muscle aches), headache, diarrhea, and chills. Fever was the most frequently reported symptom, but it was often absent, especially in the elderly and immunosuppressed patients. Cough, breathlessness, and diarrhea were present in the first or second week of illness. Most patients develop pneumonia, and severe cases often evolve rapidly, requiring critical care in an ICU.

It is important to note that many clinical features of COVID-19 are similar to that of SARS. Both diseases have a median incubation period of about five days. Also, old age and comorbidities are common risk factors for severe disease outcomes, including death, ICU admission, and mechanical ventilation.

The progression for patients with severe diseases follows a similar pattern in both infections. Lung abnormalities on chest CT scans show the greatest severity, approximately ten days after the initial onset of symptoms. However, SARS is considered more deadly than COVID-19.

The *British Medical Journal* reported WHO's director-general, Tedros Adhanom Ghebreyesus, saying that more than 80% of patients with COVID-19 have *mild disease and will recover.* However, it is fatal in 2% of reported cases. In comparison, the 2003 outbreak of SARS had a case fatality rate of around 10% (8,098 cases and 774 deaths). Yet, COVID-19 appears to be much more infectious than SARS.

Asymptomatic people are thought to be largely responsible for spreading COVID-19. A recent article published in the *Journal of Emerging Infectious Diseases* has estimated R_0 in COVID-19 to be as high as 5.7, much greater than SARS, which is supposed to vary between 2 and 4. The highly infectious nature of COVID-19 has resulted in its global spread within a few months of the outbreak.

However, despite being a fatal infectious disease, the SARS outbreak ended with the implementation of appropriate and effective infection control practices; no new cases have been reported since 2004. Unfortunately, there is still no effective vaccine against it, and treatment remains mainly supportive care.

One of the few lessons we can derive from the SARS epidemic is that it gave the world a clear warning that should have alerted public healthcare professionals and policymakers to the glaring possibility of a similar future outbreak. With the wisdom of hindsight, we can view it as a prelude to the COVID-19 pandemic due to the similarities in features. SARS has demonstrated global vulnerability to infectious diseases. More significantly, it had forewarned the world to prepare for the possibility of worse epidemics.

Swine Flu

The Swine Flu, which started in 2009, spread across the globe to cause between 151,700 and 575,500 deaths. It was first detected in the United States in April 2009, spreading rapidly across America and the rest of the world. The pathogen responsible for it was a novel influenza A (H1N1) virus, and it comprised a new combination of influenza genes previously unseen in humans or animals.

A few people had immunity against it since it was a newly-identified virus. However, about one-third of people older than 60 had antibodies against the virus, most probably from exposure to an older mutant of the H1N1 virus much earlier in their lives. The disease affected mainly children, young adults, and middle-aged adults. The clinical manifestations, such as fever, cough, runny nose, sore throat, body aches, headaches, fatigue, and more, were just as found in another typical seasonal influenza. Also, the critical diagnosis observed for severely ill patients was pneumonia, complicated by respiratory failure, which was the main cause of death from swine flu worldwide.

By May 7, 2009, the disease had spread rapidly, with the number of confirmed cases rising to 2,099. As a result, the WHO declared the 2009 H1N1 influenza a pandemic on June 11, 2009. From April 12, 2009, to April 10, 2010, the CDC estimated that there were 60.8 million cases, 274,304 hospitalizations, and 12,469 deaths in the United States due to the (H1N1) virus.

Globally, 80% of (H1N1) pdm09 virus–related deaths were estimated to have occurred in people younger than 65. On August 10, 2010, the WHO declared an end to the global 2009 H1N1 influenza pandemic, although the (H1N1) pdm09 virus remains in circulation as a seasonal flu virus, causing illness, hospitalization, and deaths yearly.

MERS

MERS, another coronavirus-induced epidemic, afflicted the world in 2012. It was an illness caused by a coronavirus named Middle East Respiratory Syndrome Coronavirus (MERS-CoV). The MERS epidemic, which started in the Arabian Peninsula in 2012, rapidly spread over 22 countries, leaving 659 deaths behind.

2,494 laboratory-confirmed cases of infection with MERS-CoV were reported, and 858 MERS-CoV-associated deaths occurred across 27 countries, according to the WHO. The first case of MERS was reported in Saudi Arabia in September 2012, although retrospective investigations revealed that the first-known cases of MERS occurred in Jordan in April 2012. According to the CDC, all cases of MERS have been linked to travel to, or residence in countries in and near the Arabian Peninsula. The largest known outbreak of MERS outside the Arabian Peninsula occurred in the Republic of Korea in 2015. The outbreak was associated with a traveler returning from the Arabian Peninsula.

MERS-CoV affects humans from direct or indirect contact with the *dromedary*, the one-humped, domesticated camel found in the Arabian Peninsula. The clinical features of MERS-CoV infection range from asymptomatic infection to acute upper respiratory illness and rapidly progressive pneumonia, respiratory failure, septic shock, and multi-organ failure resulting in death. The common features are fever, dry cough, breathlessness, sore throat, fatigue, and severe headache.

The median incubation period is approximately five days, ranging from 2 to 14 days. Most cases are adults; however, children and adults of all ages have been infected. In addition, most hospitalized MERS-CoV patients had chronic comorbidities, as seen in COVID-19 cases. In critically ill patients, the median time from onset to ICU admission is approximately five days, and the median time from onset to death is approximately 12 days.

According to the CDC, the case fatality proportion is approximately 35%, making it one of the highest among recent epidemics. MERS-CoV has spread from ill people to others through close contact, such as caring for or living with an infected person. There is no specific treatment for MERS; treatment is supportive care and the implementation of control and preventive measures.

There is no sustained human-to-human transmission of MERS anywhere in the world. However, limited unsustained human-to-human transmission in healthcare facilities remains an important feature of the virus. *One should note that the three successive coronavirus pandemics of the current century have caused nearly endless death and debility. While the three viruses are similar in clinical manifestation, their infectivity and virulence vastly differ.*

Ebola Virus Disease

Ebola Virus Disease (EVD) was first discovered in 1976 when two outbreaks of fatal hemorrhagic fever occurred in different parts of Central Africa. The first outbreak occurred in the Democratic Republic of Congo, formerly Zaire, in a village near the Ebola River, which gave the virus its name. The second outbreak occurred in South Sudan, approximately 850 km from the first location.

Scientists later discovered that two genetically distinct viruses, *Zaire ebolavirus and Sudan ebolavirus, caused* both outbreaks. *In addition,* African fruit bats are likely involved in the spread of the Ebola virus.

On March 23, 2014, the WHO reported cases of EVD in the forested rural region of southeastern Guinea. This 2014–2016 Ebola outbreak in West Africa spread to urban areas and across borders within weeks and became a global epidemic within months. On August 8, 2014, the WHO declared it a Public Health Emergency of International Concern (PHEIC), a terminology reserved only for epidemiological events with a risk of potential international spread.

Throughout the epidemic, EVD spread to seven more countries Italy, Mali, Nigeria, Senegal, Spain, the United Kingdom, and the United States. Later, secondary infection, mainly in healthcare facilities, occurred in Italy, Mali, Nigeria, and the United States. Finally, the West Africa Ebola virus epidemic started in 2013, spread to over ten countries, and led to 11,323 deaths.

With EVD, there was generally an abrupt onset of fever and symptoms, typically eight to twelve days after exposure. Fever, fatigue, vomiting, diarrhea, and anorexia were the most common symptoms of the 2014 West African outbreak. In addition, an extreme fluid loss from the gastrointestinal tract resulted in hypovolemia and metabolic anomalies, including hyponatremia, hypokalemia and hypocalcemia, shock, and multi-organ failure.

Bleeding was not universally present but could manifest later as petechiae, ecchymosis, or mucosal hemorrhage. However, frank hemorrhage was less common. The CDC reported that patients with fatal disease usually developed more severe clinical signs in the early infective phase and died between day six and sixteen of complications, mainly from septic shock and multi-organ failure.

The case fatality proportion among patients with a known outcome in Guinea, Liberia, and Sierra Leone was 70%. The US FDA (United States Food and Drug Administration) approved the Ebola vaccine rVSV-ZEBOV on December 19, 2019; the vaccine is a single-dose vaccine regimen that is safe and protective against only the Zaire ebolavirus specie of ebolavirus. However, it may not provide 100% protection for everybody.

Zika Virus

A mosquito-borne virus, named Zika virus, was identified in monkeys in a Zika forest near Entebbe, Uganda, in 1947. Again, in 1952, it was identified in humans in Uganda and the United Republic of Tanzania. The first Zika virus outbreak was reported in the Island of Yap, the Federated States of Micronesia, located northeast of Papua New Guinea, in 2007, where patients developed fever, rashes, conjunctivitis, and joint pain. This was followed by a large outbreak of Zika virus infection in French Polynesia in 2013 and other countries and territories in the Pacific. In addition, a large outbreak of Zika virus disease was reported in Brazil in March 2015.

Zika virus is mainly transmitted by the bite of an infected Aedes aegypti. This mosquito also transmits dengue, chikungunya, and yellow fever. It usually bites during the day, peaking in the early morning and late afternoon. It is also transmitted from mother to fetus during pregnancy, through sexual contact, transfusions of blood and blood products, and organ transplantation.

In October 2015, an increase in newborns with microcephaly was reported in Pernambuco, Brazil. Microcephaly is a congenital disability in which a baby's head is smaller than expected compared to babies of the same sex and age. As a result, babies with microcephaly often have smaller brains that might not have embryologically developed properly.

By 2015, 4,180 cases of suspected microcephaly had been reported. Outbreaks soon appeared throughout America, Africa, and other world regions. To date, 86 countries and territories have reported evidence of mosquito-transmitted Zika infection. On February 1, 2016, the WHO declared a Public Health Emergency of International Concern (PHEIC) concerning clusters of microcephaly cases and neurological disorders in some areas that Zika virus affected.

The incubation period of Zika virus disease is about three to fourteen days. After that, most infected people do not exhibit any symptoms, while those with symptoms usually complain of fever, rash, muscle and joint pain, conjunctivitis, headache, and malaise, lasting for two to seven days.

The most significant complication of Zika virus disease is Guillain-Barré syndrome and microcephaly. In Guillain-Barré syndrome, there is paralysis of the legs and arms due to damage to nerve fibers. In addition, Zika virus infection during pregnancy causes microcephaly and other congenital abnormalities in the developing fetus, stillbirth, and preterm birth. Unfortunately, there is no vaccine or treatment for the Zika virus disease; protection from mosquito bites is the main prevention method.

We can rationally conclude that all these epidemics and pandemics exposed our vulnerability to biological pathogens, but more importantly, they have demonstrated appalling inadequacies in our healthcare delivery systems worldwide. The pertinent question then becomes *how can we effectively curtail the human and economic damage that pandemics cause, as we are still far from preventing them in the future—an outbreak that will always possess the potential to turn into a pandemic within a relatively short period?* The answer to the question is clear; we will need careful planning, deliberate allocation of sufficient funds for research, and a commitment to strengthening our public health infrastructure.

Pandemics tend to spread rapidly across international borders. Therefore, the global spread is associated with international travel. Although international control measures are under the guidance of the WHO and implemented to limit the spread and damage, the implementation of those measures and their effectiveness will vary from one country to another. Ultimately, studying the spread and global response to pandemics is prudent to limit their consequences.

CHAPTER THREE

ORIGIN AND EARLY SPREAD OF COVID-19

The COVID-19 pandemic has spread globally, leaving millions paralyzed with severe illnesses and a harvest of deaths in the worst situations. The international response spearheaded by the WHO to halt its spread was, and remains, remarkable. Still, it spread like wildfire from its source in Wuhan, China, to nearly every country on Earth, bringing along untold death and debility. One way or the other, almost every person on Earth has been affected.

The most herculean of efforts by governments all over the world failed to stop the coronavirus; entire cities were under severe lockdown; international borders were closed; travel bans were imposed at local, national, and international levels; markets, schools, universities, and shopping complexes were shut down; quarantine and self-isolation were also advised to stop the spread of COVID-19.

The virus triggered a great global crisis that compelled the WHO to aggressively provide technical guidance to the government and key stakeholders in the global response to community spread. The

WHO continues to offer regular updates on the global situation on COVID-19 while coordinating with governments and various stakeholders to prepare for public health interventions to contain the pandemic.

The Beginning of a Tragic Story

Wuhan, the capital of Hubei Province in China, is officially reported to be where the coronavirus first appeared and infected its first victim. The Chinese authorities, however, are not entirely certain of this place of origin, resulting in confusion and controversy about the origin of COVID-19. However, multiple media houses have linked the Huanan Seafood Market at Wuhan with the virus's origin.

Many media houses also reported on several conspiracy theories about its origin. For example, many considered it a man-made biological weapon created in a laboratory in China. However, the famous Nature Medicine journal published an article that refuted the conspiracy theory on March 17, 2020, saying that it is impossible to develop that sort of virus artificially.

First Reported Case of COVID-19

A lady named W.G, according to the Wall Street Journal, is most likely the first reported case of COVID-19. These were her words, *'I felt a bit tired, but not as tired as previous years. Every winter, I always suffer from the flu. So, I thought it was the flu."* When she fell ill in December 2019 while selling shrimps at the Huanan Seafood Market, she went to a local clinic, where she received treatment but did not recover.

Afterward, she visited Wuhan Union Hospital on December 16, where she learned that several other people from the Huanan Seafood Market had visited the hospital with similar symptoms. She was placed under quarantine and recovered later, but many others had come down with cough and fever by the time she recovered. However, she was among the first 27 patients to test positive for COVID-19 and one of 24 cases with direct links to Huanan Seafood Market.

Yet, controversy surrounded this early phase. A study published in the *Lancet*, and reported by the BBC, claimed that the first case, identified on December 1, 2019, was an older man with Alzheimer's disease. But, the South China Morning Post reported that the first case could be traced back to November 17, 2019.

The Story of Dr. Li Wenliang

By the last week of December 2019, the victims had piled up; and the epidemic's severity stared doctors in Wuhan fully in the face. The patients came to the hospital complaining of cough and high fever. The chest X-rays revealed a new type of pneumonia, but many patients could not be linked directly to the seafood market. Moreover, previous experience with SARS made the doctors wonder whether they saw a similar incidence.

Suddenly, a whistleblower called Dr. Li Wenliang, who was an ophthalmologist, arose from among the ranks of the doctors. He had warned his colleagues about the possibility of an outbreak of an illness that resembled SARS and urged his colleagues to start wearing protective gear. Nevertheless, Wuhan Municipal Health Commission declared that investigations did not find any obvious human-to-human transmission and no medical staff infection, which ended the story on December 31, 2019.

The story took another dramatic turn by the first week of January 2020. Wuhan Public Security Bureau severely reprimanded Dr. Li Wenliang for *spreading rumors*. On January 3, 2020, he was forced to sign a statement acknowledging that misdemeanor; and he signed an undertaking to desist from any unlawful act, especially concerning his claims of an imminent epidemic.

Tragically, on January 10, Dr. Li developed a cough and had a fever the next day. He was diagnosed with COVID-19 on January 30. By February 7, Dr. Li succumbed to severe illness. He breathed his last, triggering unprecedented online grief and rage against the Chinese government. The tight-lipped Communist Party of China was compelled to offer a solemn apology to Dr. Li's family. At the same time, two police officers were punished for their role in the initial incident.

China Between November 17, 2019, and March 31, 2020: A Plethora of Cases

By the time Dr. Li died, the coronavirus had started spreading widely. According to the South China Morning Post, from November 17 to December 15, 2019, the total number of infections stood at 27; by December 20, the total number of confirmed cases had reached 60.

Dr. Zhang Jixian, a doctor at *Hubei Provincial Hospital of Integrated Chinese and Western Medicine,* examined an elderly couple and found them to have unusual pneumonia. She advised their son to see her; she investigated him and diagnosed him with the same disease. Her experience in the management of SARS led her to the convincing belief that it was a new virus. On December 27, she informed China's health authorities that a novel coronavirus caused the mysterious disease.

By December 31, 2019, the number of confirmed cases had risen to 266, while they rose to 381 by January 1, 2020. By February 1, 2020, the number of confirmed cases in China was more than 11,000; it rose to 82,545 by March 31, 2020, with a death toll of 3,314. On May 15, 2020, the number of confirmed cases in China totaled 84,469, and the death toll climbed to 4,644.

On January 1, 2020, the seafood market was closed; and workers in hazmat suits collected samples. However, the traders were kept in the dark and asked to leave the market. By January 20, the virus targeted the major cities of Beijing, Shanghai, and Shenzhen.

The Discovery of SARS-CoV-2

The WHO China Joint Mission presented a report that shed light on the mysterious disease while conducting an assessment from February 16 to 24, 2020. The mission reported that a new strain of coronavirus, later renamed *SARS-CoV-2,* caused the disease, and it had affected 75,465 people in China by February 20, 2020.

The median age among 55,924 laboratory-confirmed cases was 51 years, with the majority of cases (77.8%) between ages 30 and 69. Among the reported cases, 51.1% were male, 77.0% were from Hubei, and 21.6% were farmers or laborers by occupation. The Chinese government took stringent measures to control the spread of the disease, including banning public gatherings, closing schools and malls, and locking down Wuhan.

Yet, all these measures failed to stop the spread of the coronavirus. By the second half of January 2020, the virus had crossed into the neighboring countries of South Korea and Japan and finally arrived in Europe and the United States. The WHO's situation report dated January 21, 2020, recorded two cases of COVID-19 in Thailand and one case in Japan and South Korea each; but that was just the beginning of a huge global tragedy.

CHAPTER FOUR

THE GLOBAL RESPONSE TO COVID-19

Declaration of a Public Health Emergency of International Concern

A Public Health Emergency of International Concern (PHEIC) is the loudest alarm the WHO Director-General can sound. The International Health Regulations (IHR) (2005) mandates that the WHO Director-General considers the advice of an Emergency Committee convened for the purpose and drawn from a roster of experts that the WHO maintains when determining whether an event constitutes a PHEIC. In addition, the affected state is invited to present its views to the Emergency Committee.

The WHO Director-General has the final authority to declare if a PHEIC is recommended, considering all information. The WHO IHR Emergency Committee meeting to discuss the outbreak on January 22nd–23rd, 2020, was split on whether to recommend that the outbreak be declared a PHEIC. The Committee met the following week again when the Director-General returned from a mission to China.

Following the Committee's recommendation, the WHO Director-General declared that the outbreak constituted a PHEIC on January 30, 2020. At that time, there were 98 cases in 18 countries outside China. However, the Emergency Committee specified that no travel restrictions were recommended, based on the information available.

As COVID-19 spread in February of 2020 and there was an apparent lack of understanding that declaring a PHEIC was to sound the loudest possible alarm, there was increased clamor for WHO to describe the situation as a pandemic. Eventually, the WHO characterized COVID-19 as a global pandemic, stating that it was alarmed by both the spread of the disease and the level of inaction in response.

By January 30, 2020, it should have been clear to all countries from the declaration of the PHEIC that COVID-19 represented a serious threat. China had reported upwards of 20,000 confirmed or suspected cases and 170 deaths. Moreover, the number of countries where the virus had spread and where local transmission was occurring was growing by the day.

However, only a minority of countries set comprehensive and coordinated COVID-19 protection and response measures in motion; a handful did before seeing a confirmed case, and the remaining countries did once cases had arrived. Therefore, we must ask ourselves, 'Why did the PHEIC declaration not spur more action when the looming threat was evident?'

Moreover, after a lazy start to the global response in January 2020, it was clear that countries would need a full-scale response by the end of that month. It was obvious that February 2020 was already a lost month when steps could and should have been taken to curtail the epidemic and prevent the pandemic. Two things caused the failure of most countries to respond in February

2020: First, they neither appreciated the threat nor knew how to respond. Second, a *wait-and-see* approach appeared to be a less-costly and less-consequential choice than a concerted public health response.

Jumping Borders: COVID-19 Arrives in South Korea

About a month after COVID-19 was detected in China, on January 20, 2020, healthcare workers confirmed the first case of COVID-19 in South Korea. In its fight against COVID-19, South Korea established a very effective program to contain the epidemic from spreading further. Moreover, the country keenly appreciated what had happened in Wuhan and sprang into action before the virus appeared on its shores.

In a few days, South Korea prepared a test kit and started conducting tests when there were only a few cases. The test kits were distributed in the regional health centers and used to identify symptomatic and asymptomatic cases. In addition, the country adopted innovative strategies in its aggressive response. For example, there was drive-through testing, mobile applications for contact tracing, and the isolation of infected patients to halt further spread in the community. The government also resorted to other aggressive precautionary steps to control the outbreak.

By January 21, 2020, the government had scaled quarantine and screening measures for travelers from Wuhan at entry points, including the national alert level from Blue (Level 1) to Yellow (Level 2) out of a four-level national crisis management system.

However, by February 20, there was a significant increase in the number of COVID-19 cases, largely due to a gathering of over 500 people at the Shincheonji Church of Jesus at Daegu, about 240 kilometers southeast of Seoul, where worshippers from Wuhan were also present.

A 61-year-old woman, later referred to as Case 31, was present at that gathering and later tested positive for COVID-19. She had attended two such gatherings afterward and was believed to have infected numerous people, resulting in a spike in positive cases in South Korea. In addition, a 72-year-old woman, referred to as S.M.H., was among those who attended that church service. She developed a severe phlegmy cough a few days later and was admitted to a Seognam hospital about 220 kilometers away from Daegu since there were no hospital beds in Daegu.

With the rising positive cases of COVID-19 in the country, South Korea declared the highest possible alert on February 23, 2020. By February 29, the number of cases was 3,150. A week later, on March 8, 2020, South Korea announced that the outbreak associated with Shincheonji Church totaled 4,482 infections, accounting for 62.8% of the total confirmed cases. On March 13, 2020, the number of recoveries became more than that of newly tested positive cases.

A similar incident occurred between March 1 and March 8, 2020; coronavirus infected some 46 churchgoers after they attended the River of Grace Community Church. The infections were claimed to have been caused by spraying salt water into followers' mouths in the credulous belief that this would protect them from the virus. Afterward, other church clusters of the virus appeared in the cities of Suwon, Busan, Geochang, and Bucheon.

The number of cases continued to rise; by March 31, 2020, 9,786 people were infected, with the death toll climbing to 162. According to the WHO, by April 15, 2020, the total confirmed cases were 10,591, and the total number of deaths was 225. On May 15, 2020, the total number of cases was 11,018, and the death toll had climbed to 260.

South Korea received the global community's praise for its early and rapid response. Yet, it earned its fair share of criticism, especially regarding the extensive use of tracking applications and the disclosure of private information of infected people. However, South Korea could not lower its guard. The pandemic was fast evolving, and there was no herd immunity; a vaccine was still far off.

The European Conundrum

Coronavirus arrived in Europe in late January 2020. It spread like wildfire from one country to another, taking over the European continent in a few weeks; the pandemic's epicenter quickly shifted from Asia to Europe.

There were many reasons for the rapid spread: First, few thought the virus would land on their shores. They considered it an East Asian tragedy. Second, the West, and especially the medical community in the West, had little experience dealing with epidemics of infectious diseases. The virus gave little or no time for adequate preparations: there was a lack of test kits, face masks, sanitizers, and Personal Protective Equipment (PPE). Third, they did not opt for lockdowns and banning air travel at the beginning.

The usual outcry about fundamental rights violations in a rights-conscious society could only spell doom in a raging pandemic. The results were unprecedented death and debility in Europe.

France

In late January 2020, France became the first European country to confirm cases of coronavirus infection. The first case was confirmed at Bordeaux. He was a 48-year-old French citizen who arrived in France from China. France also recorded the first COVID-19 death in Europe. He was an 80-year-old Chinese tourist who arrived in France on January 16 and was admitted to a hospital, but he died on February 14, becoming the first fatality outside Asia.

A few days later, a 60-year-old school teacher became the first French national to die after developing COVID-19. Meanwhile, on February 8, 2020, French Health Minister, Monsieur Buzyn, confirmed five new cases from a group of holidaymakers in Les Contamines-Montjoie, Haute-Savoie. They contracted the infection from a British national who had attended a conference in Singapore a few days earlier.

Like South Korea, at least half of about 2500 people who attended the annual Open Door Church assembly from February 17 to February 24 in Mulhouse were believed to have contracted the virus. They became one of Europe's largest regional clusters of infections, quickly spreading across and eventually out of France. By late February, multiple cases had appeared in France, notably within three new clusters in Oise, Haute-Savoie, and Morbihan.

Coincidentally, France's first municipal elections took place on March 15, 2020, against the raging epidemic. The election, with a voter turnout of only 45%, was justified because democratic values are the essence of French life. For hygienic reasons, French nationals were supposed to bring their ballpoint pen to fill out the ballot paper. However, balancing democratic values and public hygiene made it difficult for French nationals to appear before the ballot box. Fear, anxiety, and despair became the order of the day in France.

A French citizen Monsieur C.B., aged 29, who developed COVID-19, said, *"The tests are expensive, and they are kept only for complex cases."* He was married and had a 2-year-old son. He was confined to his home in Paris. Fortunately, he improved gradually.

COVID-19 posed various other challenges. But, for a French

homemaker and mother of three children, Madame D.K., the most difficult challenge was that she was faced with the continued education of her children while confined at home.

By March 1, 2020, the number of confirmed cases in France rose to over 100, with two deaths, while the total number of cases jumped to 10,877 with 372 total fatalities by March 19, 2020. By the end of March, the virus had claimed the lives of 3,017 people while infecting 43,977 others. The number of infected people and deaths rose sharply after that; by April 15, 2020, the total number of cases and death toll became 102,533 and 15,708, respectively.

On May 15, 2020, the total number of cases had increased to 139,152, with 27,378 deaths. The BBC once reported that a patient treated in a hospital near Paris on December 27, 2019, for suspected pneumonia, had the novel coronavirus. This means that the virus may have arrived in Europe far earlier than previously believed. A swab taken at the time was tested and came back positive for COVID-19.

The French health ministry announced that the government was obtaining confirmation on the case and that it would consider further investigations if they proved necessary. Nevertheless, the virus spread and ultimately resulted in the deaths of thousands of French people. On March 16, President Macron announced a 15-day national lockdown, commencing midday on March 17; and on March 27, Prime Minister Édouard Philippe announced that the lockdown would be extended until April 15. France enforced a strict lockdown, fined those who ignored it, and imprisoned repeat offenders.

Italy: The Lombardy Tragedy

The novel coronavirus became overwhelming for Italy. The virus's presence was announced in Italy after two Chinese tourists were detected to be positive for COVID-19 in Rome on January 31, 2020. They were hospitalized in isolation at the Spallanzani hospital in the capital.

A week later, an Italian national who returned from Wuhan, China, was hospitalized and confirmed to have COVID-19. Next, Italian researchers looked into the unusually high number of severe pneumonia cases in Lombardy in the last quarter of 2019. Adriano Decarli, an epidemiology professor at the University of Milan, reported that hundreds of patients were hospitalized with pneumonia in Milan and Lodi between October and December 2019.

The virus had probably invaded Italy much earlier than the last week of January 2020. In mid-February 2020, a 38-year-old man in Lombardy fell ill and was treated by a doctor in Castiglione d'Adda. He did not recover and went to Codogno Hospital, where he was further treated, but no one suspected COVID-19. Later, his wife said he had met an Italian friend who returned from China on January 21. Both the husband and wife tested positive for coronavirus. Unfortunately, they had met many people and led an active social life before their diagnosis, infecting many others, including healthcare workers. The couple was hospitalized afterward, and Lombardy soon became the coronavirus hotspot in Italy.

Gradually, the number of cases increased, and the first death was reported on February 22. The Lombardy cluster was majorly responsible for the Italian coronavirus disaster. The highly contagious nature of the virus and lack of restriction on the movement of people allowed the infection to spread unchallenged. Ultimately, on March 8, Prime Minister Giuseppe Conte extended the quarantine lockdown to cover all the regions of Lombardy and 14 other northern provinces.

It is believed that just two days before the detection of the first COVID-19 case in Italy, a Champions League match between Atalanta and Valencia in February may have been the site from which the virus spread so rapidly in the country. Atalanta hosted the Spanish team on February 19 in Milan. During that month, nobody was aware that the coronavirus was circulating undetected. That football game was named *Game Zero* by the local media. Thousands of people watched the game in the stands, sitting only a few centimeters apart and then going to pubs and restaurants after the game. Unfortunately, there were also about 2500 Valencia fans in the stands for the match; 35% of Valencia's team tested positive for the novel coronavirus a few weeks later.

A journalist who covered the match became the second person to test positive for COVID-19 in Bergamo. Additionally, several Valencia fans became infected. That football match proved to be a biological bomb that exploded later in the form of devastating consequences.

Another cluster of cases appeared in Veneto, Northeast Italy. Gradually, the virus continued its deadly march across central and southern Italy, leaving an unprecedented number of casualties. In certain areas of northern Italy, the casualties were so high that hardly anyone was left to bury the dead. For example, in Lombardy, families could not give proper farewells to their dead relatives, and only a few minutes of funeral services were performed.

The tragedy of COVID-19 is that, while hospitalized, a patient could not meet their relatives because most of them were in quarantine. They died alone and were cremated with only a few dear ones around. The cemetery was so overwhelmed at Bergamo that military trucks were seen carrying dead bodies to other places for burial. There was hardly any family left untouched by the virus in certain villages and towns as grief, despair, and hopelessness was predominant. Moreover, the Italian government declared a state of emergency.

In February, 11 municipalities in northern Italy were placed under quarantine. Prime Minister Giuseppe Conte extended the quarantine to Lombardy and 14 other northern provinces on March 8, 2020; he also extended it to all of Italy the following day. On March 21, the government ordered the shutdown of all non-essential businesses and industries. Lockdown in Italy was especially strict, and the police fined anyone who ventured out of their home without a valid reason.

Spain

Spain confirmed its first case of COVID-19 on January 31, 2020. CNN reported that someone in La Gomera, one of Spain's Canary Islands, was part of a group of five people in contact with a person infected with the novel coronavirus while in Germany. On February 9, Reuters reported that a Briton tested positive for the coronavirus in Mallorca, making him the second COVID-19 patient in Spain. He was one of four members of a British family under close observation in Mallorca after coming into contact with someone in France who was subsequently diagnosed with COVID-19. Fortunately, the test results for the man's wife and two daughters were all negative.

After the famous *Game Zero* in Milan, the Valencia fans returned to Spain, ignorant of the deadly baggage they had brought back home. Unfortunately, most of them tested positive afterward. On February 27, a health department spokesperson in the Valencia region confirmed six new cases of COVID-19 in the city. Three days later, a Portuguese who had visited Valencia tested positive after returning home.

On February 24, an Italian was kept in isolation at H10 Costa

Adeje Palace Hotel, Tenerife; and the hotel was locked down. Four other people tested positive too. The total number of active cases in the Canary Islands was six—five in Tenerife and one in La Gomera. More than 700 tourists were quarantined inside the hotel, waiting for a 14-day isolation period.

A young male resident of Seville also caught the deadly virus. He was admitted to the ICU and ultimately fought off the virus. But, in relating his ordeal, he said, *"Can you imagine, you're someone who is 29, who plays sports, doesn't smoke and hardly ever goes to the doctor, and you're in intensive care with oxygen, and no one from your family is allowed to see you?"*

As the virus steadily attacked its victims, the Spanish authorities ignored the early signs. Life in Spain was business as usual. Rallies for International Women's Day on March 8 brought thousands onto the streets all over Spain, including a crowd estimated at 120,000 in Madrid. Two female cabinet ministers who attended the event later tested positive for the novel coronavirus.

The virus kept claiming its victims all over Spain, and by March 18, 2020, the total number of confirmed cases and the death toll had increased to 11,178 and 491, respectively. By March 31, 2020, the virus had infected 85,195 people and claimed 7,340 lives. The total number of confirmed cases by April 15 was 172,541, and the death toll had increased to 18,056. On May 15, 2020, the total number of confirmed cases was 229,540, and the death toll increased to 27,321.

Doctors in Spain noticed pneumonia-like cases. They quickly alerted the authorities, referencing what had happened in Italy; but their admonitions fell on deaf ears. The authorities made no arrangements to procure face masks, gloves, or PPE and made no efforts to upgrade the healthcare system to tackle the deluge of cases that would soon overwhelm the Spanish healthcare system.

The Spanish government imposed a nationwide quarantine on March 14. Then, on March 28, the government tightened its national lockdown, ordering all non-essential workers to remain at home for the next two weeks, with the faint hope of stopping the deadly march of coronavirus; it was all to no avail.

The United Kingdom of Great Britain and Northern Ireland

On January 31, England reported its first case in North Yorkshire. Two Chinese nationals from the same family who were guests at a hotel in Yorkshire tested positive for COVID-19. On the same day, a flight from Wuhan landed in the UK; and all the passengers were placed in quarantine.

The third patient was diagnosed at Brighton on February 6; he managed to infect at least 11 people. His story was a classic study of the spread of a pandemic due to globalization. First, he attended a business conference, in Singapore, with 100 attendees, including someone from Hubei, China. Later, he traveled to the French Alps to holiday at the ski resort of Contamines-Montjoie with a group of other British citizens. Then, he returned to the UK with a cough and fever.

Subsequently, on February 27, the first case was diagnosed in Northern Ireland; and another in Wales the next day. Finally, on February 28, a Surrey resident was diagnosed with the coronavirus, although he did not have a recent travel history. He was transferred to a specialist National Health Service (NHS) infection center at Guy's and St Thomas' Hospital in central London.

By March 2020, the virus had affected several areas of the UK and started claiming its victims. The first to succumb was a 70-year-old woman in Berkshire. She was being treated at the

Royal Berkshire Hospital in Reading and believed to have caught the virus in the UK. The next to die was an 80-year-old man at Milton Keynes Hospital.

By March 21, 2020, the virus had attacked 5,018 people and killed 233. By March 31, 2020, 22,145 cases had been confirmed, with 1,408 deaths. The WHO reported that the total number of confirmed cases and the death toll had increased to 93,877 and 12,107, respectively, by April 15, 2020. Gradually, the number of infections and the death toll increased all over the UK. On May 15, 2020, the total number of cases increased to 233,155, and the death toll climbed to 33,614.

The response of the British government was different from that of most other countries. In contrast, the UK government elected not to shut down large gatherings or introduce stringent social-distancing measures. Rather, a plan was implemented to suppress the virus through gradual restrictions. That strategy was to build herd immunity. The government believed it to be a mild infection that would automatically raise the level of immunity to stamp out the virus if it were to infect a significant proportion of the population.

Imperial College, London, argued that this strategy would overwhelm the healthcare facility, finally leading to a U-turn by the government, which introduced new social-distancing measures. On March 23, the prime minister ordered the closure of pubs, gyms, and cinemas.

The Coronavirus Act 2020 received Royal Assent on March 25, having been fast-tracked through Parliament in a four days record of legislative seating. Meanwhile, Prime Minister Boris Johnson and the Prince of Wales tested positive for the deadly virus. Johnson used his case as a teachable moment for the country, appealing to people to work from home and comply with the more drastic social-distancing measures he had implemented. He eventually spent three nights in ICU at St Thomas' Hospital.

Germany

On January 27, the first case of COVID-19 was reported near Munich, Bavaria, Germany. The patient had no history of travel to China, but there was contact with a Chinese guest at the company where he was employed. The next cases were from German car parts supplier, Webasto's headquarters in Stockdorf, Bavaria. The company reported that a Chinese employee from Shanghai had tested positive for the virus upon his return to China following his visit to the company headquarters. Many cases later developed and were linked with Webasto.

By March 1, 2020, the number of confirmed cases was 129 in Germany, although no deaths were reported. However, on March 8, an 83-year-old resident of the St. Nikolaus Home for The Elderly in Wurzburg was diagnosed with COVID-19 and admitted to a hospital. He died four days later, becoming the first casualty of COVID-19 in Bavaria.

By March 27, ten more residents of the same home died; and several residents and employees tested positive. The Wurzburg home became the virus hotspot of COVID-19 death. Many residents at old homes all over Germany suffered the same fate. Virtually all of these senior citizens also had serious comorbidities, including dementia.

Another cluster of patients appeared at Baden-Wurttemberg. On February 25, a 25-year-old man from the district of Goppingen, about 40 kilometers away from Stuttgart, recently returned from Milan, Italy and fell ill with flu-like symptoms. He contacted the local health department and was isolated and treated in a clinic.

The next day, a 24-year-old girl and her father in nearby Tubingen were also diagnosed with COVID-19. The girl was a friend of the 25-year-old man and most likely contracted the disease from him. After that, the cases steadily piled up in Baden-Wurttemberg. Most of these patients had contact with Italy, to which they had either traveled or had come into contact with someone with a history of travel to Italy.

The virus attacked North Rhine-Westphalia in its initial march through Germany. On February 25, a 47-year-old man and his wife tested positive in Erkelenz, Heinsberg. The couple had attended a carnival in Langbroich village, where they proceeded on a short visit to the Netherlands. The man became seriously ill and was placed on ventilator support.

The virus spread through several areas from these initial cases, and the case count started increasing. For example, one of the infected people from the cluster in North Rhine-Westphalia was a doctor at Maria Hilf hospital in Monchengladbach. After the case was identified, at least 35 people, including 15 of the doctor's patients, were medically examined. By March 16, Germany confirmed at least 6,600 cases of COVID-19. Of them, 2,400 were in North Rhine-Westphalia, more than 1,000 in Bavaria, and more than 900 in Baden-Wuerttemberg.

By March 19, 2020, the total number of cases and fatalities in Germany were 10,999 and 20, respectively; by March 31, 2020, the virus had infected 61,913 people and claimed the lives of 583 people. As a result, the total number of confirmed cases increased to 127,584, and the death toll also increased to 3,254 by April 15, 2020. On May 15, 2020, the total number of reported cases became 173,152, and the death toll climbed to 7,824.

Germany seemed to have taken the epidemic in its stride with a high number of cases but a low incidence of death. The mortality rate in Germany was far below its neighboring European countries, which was attributed to Germany's decision to implement widespread testing of people suspected of having the virus, as opposed to the Italian or British decision to test only symptomatic cases. The delayed arrival of the virus in Germany was probably another factor.

Additionally, most patients were treated at top-notch medical institutions, and the slow increase in the number of cases helped keep the death toll low. Germany, undoubtedly, has one of the best healthcare infrastructures in the world. By late January, the government was confident it would be able to contain the spread of the virus. Even in the middle of February, the government did not consider curtailing flights to and from China.

By late February, the government started taking steps, like closing schools and malls and regulating air and sea travel. On March 2, the Robert Koch Institute raised its threat level for Germany to moderate, and the European Centre for Disease Prevention and Control raised its threat level for Europe from moderate to high.

The American Disaster

On January 19, 2020, a 35-year-old man was presented at a clinic in Snohomish County, Washington, complaining of cough and fever. He had a history of traveling to Wuhan, China. After collecting a nasopharyngeal swab from him, he was asked to isolate himself at home, with active monitoring by the local health department. On January 20, 2020, the US CDC confirmed that the swabs tested positive for COVID-19.

A woman in her 60s became the second case of COVID-19 in the US when she returned to Chicago on January 13 after caring for her father in Wuhan, China. A third case was confirmed a day later in Orange County, California. The significance of this trend is that, in just a few days, the virus was already in areas separated by hundreds of miles. On January 30, the first case of person-to-person transmission was confirmed. The husband of the second case did not travel to Wuhan but had frequent close contact with his wife after she returned from Wuhan. He was admitted eight days later and tested positive for SARS-CoV-2. Fortunately, both recovered from the illness.

With each passing day, new cases were detected all over the US, as the virus spread rapidly and at alarming speed and ferocity. Finally, the virus claimed its first fatality, a man in his 50s, on February 29, at Evergreen Health Medical Center in Kirkland, Washington, a suburb of Seattle.

On March 1, 2020, Governor Andrew Cuomo of New York announced the state's first reported case of COVID-19—a woman in her late 30s. She was a healthcare worker who contracted the virus while traveling in Iran and was self-isolating at home in New York City. She experienced mild respiratory symptoms and went to Mount Sinai Hospital, where she tested positive for the novel coronavirus.

New York City Health Commissioner, Oxiris Barbot, said the woman's neighborhood was not at increased risk. *"We know that there's currently no indication that it's easy to transmit by casual contact. There's no need to do anything special in the community,"* Barbot said, but she was fatally wrong. The US was caught unawares and unprepared. Throughout January, Mr. Donald Trump, 45th president of the US, repeatedly played down the seriousness of the virus and focused on other issues.

According to CNN, Donald Trump downplayed the coronavirus threat multiple times in the early months. The Central Intelligence Agency (CIA) issued several warnings, which he ignored. He even ignored the warnings in the Presidential Daily Brief, a sensitive report produced before dawn each day and designed to call the president's attention to the most significant global developments and security threats.

The virus spread throughout New York and attacked more people with little defense against it. On March 14, it claimed its first victim in New York. She was an 82-year-old woman with chronic lung disease who died after hospitalization. A Bloomberg News editor was infected in the second week of March. She had a dry cough, high-grade fever, vomiting, sore throat, severe muscle pain, and chills. A few days later, she became delusional and was rushed to the hospital. Her test turned out to be positive. A chest X-ray revealed that she had pneumonia. Yet, she fought and overcame the disease. A few days later, her roommate developed similar symptoms.

By mid-March, mitigation measures expanded in New York, with a transition to online classes for universities and colleges. In addition, the first semi-containment zone was announced in New York, and on March 20, Governor Andrew Cuomo issued a state-wide order that all non-essential workers must stay at home. On the same day, coronavirus cases in New York exceeded 7,000.

By April 3, 2020, at least 102,870 confirmed coronavirus cases were discovered in New York, including more than 57,159 in New York City alone. Within three months, 2,935 people with COVID-19 died in the state, about 38% of confirmed cases in the US. The high density of the population of New York, the highest rates of public transportation ridership, and cold weather have been the probable reasons New York became the hotspot of COVID-19 in the US.

Although the US government failed to appreciate the gravity of the situation, it later moved swiftly to outpace the virus and save its citizens. On January 29, 2020, President Trump established the White House Coronavirus Task Force to coordinate and oversee efforts to *monitor, prevent, contain, and mitigate the spread* of COVID-19 in the United States. In addition, a series of travel restrictions was announced between late January and mid-March; and Foreign nationals who had traveled through China, Iran, and 26 European countries that comprise the Schengen Zone were denied entry.

Later, the ban included travel to the UK and Ireland. American nationals who had traveled to those countries also underwent quarantine after the health screening. In early March, the US CDC advised Americans against nonessential travel to countries affected by COVID-19. By March 21, the governors of New York, California, and other states, had ordered most businesses to close and for people to remain indoors. Throughout March, several states, cities, and county governments imposed stay-at-home quarantines on their populations to halt the spread of COVID-19.

By March 19, 2020, the virus had infected 10,442 people and claimed 150 lives; by March 25, 2020, the confirmed cases and death toll had risen to 51,914 and 673, respectively. As of March 31, 2020, 140,640 people had been infected, and the case fatality had risen to 2398; by April 15, 2020, the total number of confirmed cases and deaths had increased to 578,268 and 23,476, respectively. On May 15, 2020, there were 1,361,522 cases, and the death toll had climbed to 82,119.

By the time this book went to press, the death toll from COVID-19 in the US had crossed the 1,000,000 mark and still counting.

Canada's Woes

Canada was not caught unprepared. On January 15, 2020, the Canadian Public Health Agency activated its Emergency

Operation Centre to effectuate Canada's response to COVID-19 and implemented screening protocols at airports for travelers returning from China on January 22, 2020.

The first case related to travel to Wuhan was confirmed in Canada on January 25, 2020. A 56-year-old man in Toronto, Ontario, Canada, developed fever and dry cough a day after returning from a three-month visit to Wuhan, China. He had a fever, cough, and bilateral pneumonia. Fortunately, he recovered and was discharged afterward.

On February 20, 2020, Canada confirmed its first case related to travel outside mainland China. The 77-year-old man had other comorbidities, becoming the first Canadian fatality of the virus. He died at the Royal Victoria Regional Health Centre (RVH) in Barrie, Ontario, on March 11.

Sophie Grégoire Trudeau, the wife of Canadian Prime Minister Justin Trudeau, also tested positive for coronavirus after returning from a trip to London. By March 31, 2020, the virus had infected 6,317 people and claimed 66 lives. According to the WHO, by April 15, 2020, the total number of confirmed cases and the death toll in Canada had increased to 26,146 and 823, respectively.

On March 16, 2020, Canada advised incoming travelers to self-isolate for 14 days. In addition, the country took further steps to prevent the epidemic from spreading by banning the entry of foreign nationals from all countries except the United States. However, the Canada–US land border was closed to all non-essential travel. Yet, the government's best efforts did not stop the virus in its tracks. By May 15, 2020, there were 72,536 confirmed cases with 5,337 deaths.

The Australian Fiasco

If Australia thought that because it is far-flung from the other continents, it would be spared the scourge of the novel coronavirus; it was despairingly mistaken. The world is now a global village, and Australia is no longer such a remote region. The lethal virus traveled across the sea to reach the coast of Australia.

ABC News reported the detection of the first case in the Australian state of Victoria on January 25, 2020. The infected individual was a Chinese citizen who had recently visited Wuhan. The man was admitted to a Melbourne hospital. On that day, three men aged between 30 and 60 tested positive for coronavirus, all in New South Wales. A couple of them had recently visited Wuhan, while the third had direct contact with a confirmed case from Wuhan while in China. They were all kept in isolation in New South Wales hospitals. By the end of January 2020, several other cases had been diagnosed, and almost all of them had a history of travel to China or a recent history of contact with someone with a recent history of travel to China.

In February 2020, the cases slowly started multiplying; and by late February, 24 Australians were found to have been infected on the Diamond Princess cruise ship. Then, on March 1, a 78-year-old man evacuated from that cruise ship died of COVID-19 at a hospital in Perth. He became the first Australian casualty.

The next day, a 53-year-old healthcare worker, who did not have any travel history, became the first case to be infected through community transmission. In early March 2020, a woman flew to Australia from London to attend a friend's wedding; but she couldn't attend the wedding as she contracted the virus. She spent the next 13 days in a Brisbane hospital. Her condition deteriorated, but she recovered afterward.

By March 31, 2020, the virus had infected 4,359 people and

claimed the lives of 18 people. World-famous Hollywood actor Tom Hanks and his wife, Rita Wilson, were infected in mid-March. They returned to Los Angeles after spending more than two weeks in quarantine in Australia. Hanks wrote on Instagram, *"Good News: One week after testing positive, in self-isolation, the symptoms are much the same. No fever… . We are all in this together. Flatten the curve."*

In the third week of March, the University of Queensland postponed lectures after three students tested positive for the virus. Meanwhile, many infected people disembarked from several cruise ships docked at various ports in Australia. The cases continued to rise throughout Australia, even as the government tried its best to contain the pandemic.

The Australian government implemented social distancing, a ban on public gatherings, border restrictions, and the closure of cinemas. Yet, by April 15, 2020, the total number of confirmed cases had risen to 6,416, with 61 deaths. On May 15, 2020, there were 6,989 cases, and the death toll had climbed to 98.

COVID-19 has claimed thousands of lives in several parts of Europe, the US, and China. Despite having the best possible healthcare infrastructure in the world, none of these countries could control the surge in cases. The infection started in the cold regions and spread rapidly to other countries with similar weather types. It then remained to be seen how it might spread in the hot and humid regions of the world.

The pandemic is spreading rapidly with no end in sight. The number of cases and death tolls are mounting with every passing day. The unaffected areas are rapidly turning into new playgrounds for this deadly virus. There is no vaccine and no therapeutic measures to control the spread.

David Nabarro, the WHO's special envoy for COVID-19, has

said that the coronavirus will not disappear, so we must learn to live with it. Until an effective vaccine becomes available, non-therapeutic measures like lockdowns, travel bans, and school and college closures will play important roles in controlling the pandemic.

However, with the ease of lockdowns in several countries, people have started coming out of their homes. The more people move around, the greater the chances of catching the infection. In such a scenario, it becomes crucial to learn how to recognize a case of COVID-19 and take effective precautionary measures to prevent infection.

Besides infection prevention, it is extremely important to take care of the patients and reduce death and debility from COVID-19. Early and correct diagnoses, based on clinical features and appropriate laboratory tests followed by effective treatment, will go a long way in winning the war against COVID-19.

CHAPTER FIVE

COVID-19: CLINICAL PRESENTATION, DIAGNOSIS, TREATMENT, AND PREVENTION

COVID-19 has posed critical challenges to everyone, from frontline public health workers to other essential service workers. That is why it has become imperative that the war against COVID-19 be waged collectively and individually. That is also why it is important for everyone to gain familiarity with the symptomatology of the disease, how it is transmitted, and what treatment is available for treating a patient infected with COVID-19.

Recognizing COVID-19

Early detection of COVID-19 is crucial to timely treatment, and that can go a long way toward saving a patient's life and controlling the spread of the disease in the community. In the early stages of illness, most patients infected with the novel coronavirus show symptoms similar to the common cold or flu. Therefore, it becomes crucial to differentiate between COVID-19 and the common cold. The WHO, however, has provided guidelines for case definition.

Suspicion of COVID-19

According to WHO's guidelines, a person is a suspect case if he is:

1. A patient with acute respiratory illness (fever and, at least, one sign/symptom of respiratory disease, e.g. cough, shortness of breath) with no other etiology that fully explains the clinical presentation and a history of travel to or residence in a country or territory reporting local transmission of COVID-19 during the 14 days before the onset of symptoms.

2. A patient with any acute respiratory illness who has been in contact with a confirmed or probable COVID-19 case 14 days before the onset of symptoms.

3. A patient with a severe acute respiratory infection (fever and, at least, one sign/symptom of respiratory disease, e.g. cough, shortness of breath) who requires hospitalization and has no other etiology that fully explains the clinical presentation.

Probable Case of COVID-19

A suspect case for whom testing for COVID-19 is inconclusive becomes a probable case, *inconclusive* being the result of the test reported by the laboratory.

Confirmed Case of COVID-19

A person with laboratory confirmation of COVID-19 infection, irrespective of clinical signs and symptoms, is a confirmed case of COVID-19.

COVID-19 Contact

Because the virus is moving from one person to another within the community, anyone may be exposed to someone with COVID-19. Unfortunately, many infected persons are asymptomatic. Therefore, a healthy person becomes a contact if exposed to a patient with COVID-19. The WHO defines contact as a person who has experienced any one of the following exposures during the two days before and the 14 days after the onset of symptoms of a probable or confirmed case:

1. Face-to-face contact with a probable or confirmed case within 2 meters and for more than 15 minutes.
2. Direct physical contact with a probable or confirmed case.
3. Direct care for a patient with probable or confirmed COVID-19 cases without using proper PPE.
4. Other situations as indicated by local risk assessments.

Basic Clinical Scenarios

The first and most important step in identifying a case of COVID-19 is travel history. If a person who has traveled to countries like China, the US, Italy, France, Spain, and the UK develops cough and fever subsequently, they most likely have COVID-19. The most important next step is to keep a safe distance from such a person and inform the healthcare authorities for a proper diagnosis and treatment. History of travel was extremely important in the early months of the pandemic. However, since the pandemic has spread to every part of the world, travel history has lost some relevance.

Symptoms

A COVID-19 patient may be asymptomatic. Some develop respiratory failure, sepsis, septic shock, and multiple organ dysfunction syndromes at the other end of the spectrum. The most common symptoms are fever, dry cough, shortness of breath, body ache, and fatigue. However, fever may be associated with chills.

Less-common symptoms are anorexia, sputum production, sore throat, confusion, dizziness, headache, runny nose, chest pain, presence of blood in cough, skin rash or discoloration of fingers and toes, nascent loss of taste or smell, diarrhea, nausea/vomiting, and abdominal pain. In addition, certain neurological features like tingling or numbness in the hands and feet, dizziness, confusion, delirium, seizures, and stroke are seen in some patients.

Gastrointestinal symptoms, such as loss of appetite, nausea, vomiting, diarrhea, and abdominal pain or discomfort, might start before other symptoms, such as fever, body ache, and cough. Heart involvement, leading to myocarditis, myocardial infarction, or arrhythmias, has been reported in COVID-19 patients. Ocular symptoms like itching, redness, tearing, discharge, and foreign body sensation are also found in COVID-19 patients. In addition, conjunctivitis has been reported in some patients.

For most patients, COVID-19 remains a nuisance, like the common cold; unfortunately for some patients, it can quickly transform into a struggle between life and death. Some patients experience severe cough and describe it as, for example, "I was coughing like I was going to die." Others lose their voice due to incessant cough.

A 55-year-old resident of Washington once complained of the heaviness of the chest that it was as if an elephant was standing on his chest. Some others described their condition as *taking every breath has become a battle*. As a result, patients complain of high-grade fever and constant dry cough. Some require hospitalization, and after a few days, they develop breathlessness, leading to Acute Respiratory Distress Syndrome (ARDS). Such patients require ICU care and mechanical ventilation.

Case Studies

A 48-year-old university professor at Busan, South Korea, said he felt a very mild sore throat an*d dry cough*. On February 24, 2020, he developed breathing problems early in the morning and tested positive for coronavirus. He was later admitted to a negative-pressure room in the quarantine section of Kosin University Gospel Hospital's ICU. He received oxygen, and a CT scan of his chest was done. By now, he had developed severe chest pain. "I felt a burning pain in my chest and stomach, although I was not sure if it was because of the drugs I took or the virus. I had a slight fever and my condition fluctuated. At first, I felt as if a heavy iron plate was pressing on my chest. The stabbing pain gradually eased to the point that it felt as if someone was squeezing my chest hard," he said. After his recovery from the ordeal, he said, "I was naive and stupid to think that the outbreak was not my problem. Yes, as usual, I was stupidly overconfident." Most people were overconfident.

Shortness of breath is an early sign of danger. Often, these patients require oxygen and deteriorate rapidly. A South African patient wrote: *The best way to describe it is, when you are at a high altitude, you struggle to breathe. Within days, it fluctuates. You get chilled, and later on, you feel better. The worst for me was last week. I was really short of breath. I called my doctor*. An Italian COVID-19 survivor said, "You are there on the bed attached to the oxygen, you feel breathless and you understand it may be your last breath."

Approximately 80% of patients are present with mild illness, 14% with severe illness, and 5% with a critical illness. Severe illness is seen mainly in the presence of other comorbidities, like diabetes, hypertension, lung diseases, kidney dysfunction, and others. The most common diagnosis in patients with severe COVID-19 is severe pneumonia. Kidney involvement is frequent in COVID-19, with under 40% of cases having abnormal proteinuria upon hospital admission.

Acute kidney injury is common among critically ill patients with COVID-19, affecting approximately 20%–40% of patients admitted to intensive care. In severely ill patients, liver damage with raised liver enzymes is also reported. However, many patients were found to be asymptomatic and are thought to have extensively spread the disease in the community.

The virus affects children and young adults, though their infection rate and disease severity are much lower than those of older people. However, this should not lead to a false sense of security. Children are typically asymptomatic or present with mild symptoms like fever, cough, and sore throat. There have been several cases of Kawasaki disease, an overlap syndrome of toxic shock, and Kawasaki syndrome in children with COVID-19. Presentations include fever, conjunctivitis, a polymorphous and blanching rash, tongue involvement, and swelling of hands and feet. Neonates and infants have also been reported to have developed COVID-19 and experienced diarrhea. Pregnant women with COVID-19 have similar features to those reported for non-pregnant adults.

Immediate Personal Intervention

If a person has a fever, cough, chest pain, or breathlessness, they may have COVID-19. The majority of patients have mild symptoms and recover at home. The most important first step is to contact a healthcare provider. The CDC suggests the following steps:

- Stay at home except to get medical care.
- Take care of yourself.
- Stay in touch with your doctor.
- Avoid public transport.
- Stay away from others in a separate room with an attached toilet/bathroom.
- Monitor your symptoms, and inform your doctor.
- Wear a mask near other people or animals, including pets.
- Cover your mouth and nose while coughing or sneezing.
- Clean hands frequently with soap and water.
- Avoid sharing personal items.
- Clean all high-touch surfaces every day.

Laboratory Investigation

Molecular testing is required to confirm COVID-19 diagnosis. However, there is a shortage of diagnostic kits in almost all the countries affected by the COVID-19 pandemic. Majorly, diagnostic tests have been performed according to guidance issued by local health authorities and adhered to appropriate biosafety practices.

The Nucleic Acid Amplification Test

A nucleic acid amplification test, such as Reverse Transcriptase Polymerase Chain Reaction (RT-PCR), is used to diagnose SARS-CoV-2 infection, with confirmation by nucleic acid sequencing when necessary. In addition, a nasopharyngeal or oropharyngeal swab is collected for molecular testing, but a sample should not be taken from the nostrils or tonsils.

The WHO recommends that additional tests be done if a negative result is obtained from a patient with a high index of suspicion for COVID-19. Its guidelines recommend that two consecutive negative tests (at least one day apart) are required to exclude COVID-19. Decisions about whom to test should be based on clinical and epidemiological factors, along with guidance issued by local health authorities. The WHO recommends prioritizing people with a likelihood of infection and using appropriate PPE for specimen collection, viral swabs (sterile Dacron or rayon, not cotton), and viral transport media.

Chest X-Rays

Chest X-rays reveal pneumonia in patients experiencing severe cough with high-grade fever and breathlessness. As a result, doctors worldwide face two major challenges—first, the sheer number of patients-overwhelmed hospitals; second, sick patients need ventilator support, and there has generally been an acute shortage of ventilators worldwide.

Treatment

There is no specific treatment for COVID-19 yet. The mainstay of management is early diagnosis and supportive care to relieve symptoms. Serious patients should be admitted to the hospital, and some will require ICU care. Infection prevention protocol should be implemented as soon as possible. The WHO recommends the following:

- Patients must be kept in well-ventilated rooms.
- Each patient should be kept in their room.
- If a single room is unavailable, keep patients in a single ward 2 meters apart.
- Use protective gear and single-use or disposable equipment.
- Limit the number of healthcare workers (HCWs).

- Consider keeping patients in a negative-pressure room.
- All of these patients may require oxygen therapy.
- Fluid management is essential to prevent shock.
- Preventing and managing complications like ARDS, sepsis, and septic shock should be prioritized.
- Starting antibiotics to prevent bacterial infections and choosing antibiotics depends on the local protocol.
- Antipyretics and analgesics are used to treat fever and pain. Some authorities have advised against the use of ibuprofen. Instead, the NHS in the UK recommends paracetamol as the drug of choice.

Older patients, especially those with comorbidities, require a lot of care because they are most likely to deteriorate rapidly. At the same time, young patients who appear well may also deteriorate rapidly. It has been observed that some patients collapse suddenly and require mechanical ventilation.

Corticosteroids have been used in some patients with COVID-19. However, they are ineffective and are not recommended. Similarly, many experimental therapies have been tried on compassionate grounds, like interferon, chloroquine/hydroxychloroquine, azithromycin, lopinavir/ritonavir, and convalescent plasma, with doubtful results. Therefore, on May 1, 2020, the US Food and Drug Administration issued an emergency use authorization for the investigational antiviral drug, remdesivir, to treat suspected or laboratory-confirmed COVID-19 in adults and children hospitalized with severe disease.

Less-serious patients with risk factors for poor outcomes (i.e. over 60, presence of comorbidities) should also be prioritized for hospital admission. However, the WHO recommended that confirmed cases without any comorbidities and risk factors be treated at home due to the shortage of beds. For home care, there should be some family members to look after the patient. A healthcare worker must visit this patient regularly and keep in touch with the local health authority. Patients and their family members must strictly follow the infection prevention protocol.

If symptoms worsen, they should be transferred immediately to a well-equipped hospital. A patient requires two negative test results (on samples collected at least 24 hours apart) before he or she can be released from home isolation.

Prevention of Infection

The WHO recommended an infection prevention protocol, which should be implemented for suspected or confirmed cases of COVID-19. Every healthcare worker (HCW) must follow hand and respiratory hygiene, use appropriate PPE and practice safe waste management and sterilization of patient-care equipment. PPE effectiveness depends largely on adequate and regular supplies, adequate staff training, appropriate hand hygiene, and appropriate human behavior.

These respiratory hygiene protocols should be adhered to:

- HCWs should ensure that all patients cover their mouths and noses, cough, and sneeze into their flexed elbows.
- HCWs should offer a medical mask to suspected COVID-19 patients.
- Patients visiting the hospital or a clinic must be advised to wash their hands with soap and running water before entering the premises. In addition, every patient waiting outdoors for treatment should be asked to keep a minimum distance of two meters from other patients.
- Close contact must be avoided.
- HCWs must practice proper hand hygiene before and after examining every patient. Hand hygiene means cleaning

hands with an alcohol-based hand rub or soap and water.

- Hospital and environmental surfaces should be regularly cleaned with water and detergent, and sodium hypochlorite should be applied for disinfection.
- Doctors and nurses should wear hazmat suits while taking care of patients.

Management of Contacts

All contacts should self-isolate at home. They should stay in a separate room with an attached toilet and bathroom. Also, their clothes and utensils should be washed and cleaned separately. They should be asked to monitor their health regularly and check their temperature at least twice daily. If they develop fever and/or cough, they must visit a designated treatment center to test for COVID-19. While traveling, they must wear face masks and keep a distance of at least 2 meters from everybody.

It has been reported that some patients test positive sometime after discharge; this is disease reactivation. Such people will be a source of infection to others in the community.

CHAPTER SIX

A ROLLER COASTER OF SOCIAL AND ECONOMIC WOES

The impact of the COVID-19 pandemic on the world has brought the unanticipated destruction of lives and the livelihoods of millions of people. The harvest of deaths, the lockdown, loss of jobs, totally-swamped healthcare systems, isolation, and quarantine have all generated unprecedented anxiety and fear.

COVID-19 has succeeded in changing the way of life of millions of people across the globe. Social distancing, self-isolation, quarantine, curfew, travel bans, and closures of social amenities and facilities have disrupted normal life. Moreover, the cancellation and postponement of significant global events, including the Tokyo Olympics, left the world gasping at the sheer enormity of the potential impact of COVID-19.

Economic devastation has been inestimable, with whole city lockdowns, curfews, and the closure of merchandise outlets and industries, leading to job losses. The Guardian reported that a record 36 million Americans filed for unemployment in just two months. By April 2020, 20 million people had lost their jobs, and the unemployment rate shot up to 14.7% from just 4.4% in March 2020. Those without a college education, African-Americans, and people of Latin extraction were the most affected.

At the same time, migrant workers in India lost jobs in the cities and were forced to walk hundreds of miles to their homes in remote villages amid the lockdown. Markets worldwide crashed, wiping out millions of dollars of investors' wealth.

Mental Health

Loss of sleep, nightmares, difficulty with concentrating on menial tasks, family neglect, substance abuse, and worsening overall health have contributed to a universal decline in the world's mental health status. It could not have been otherwise, with overwhelming dread and panic setting in.

Thomas Schaefer, the finance minister of Germany's Hesse State, committed suicide after becoming *deeply worried* about how to cope with the economic fallout from the pandemic. Similarly, a suspected coronavirus patient committed suicide by jumping from the seventh floor of Safdarjung Hospital, New Delhi. As a result, anxiety and depression assumed an all-time global high. While those who could easily avail themselves of help did so, the majority remained in blissful ignorance of their condition.

The stigma attached to psychiatric disorders did not help, preventing many from seeking and obtaining proper treatment. China's experience with the SARS epidemic of 2002–2003 is instructive. The epidemic killed many and led to an economic crisis that deeply impacted people's social lives and mental health. The high death rate profoundly impacted the Chinese people, and many, including healthcare workers, developed depression and anxiety. In addition, many survivors developed post-traumatic stress disorder.

The COVID-19 pandemic is posing a similar challenge to people all over the world. People react differently to stress, depending on various factors, including socioeconomic status, accessibility to healthcare, educational background, and social and family support. The following groups are the most vulnerable to stress:

- The homeless and the poor
- Migrant workers
- Drug addicts
- Children and teenagers
- Older people with a chronic disease
- Frontline Healthcare workers
- Those already experiencing psychological dysfunction

The current mental health landscape poses a tremendous challenge to psychiatrists and other mental healthcare workers worldwide. While a visit to a psychiatrist may address certain issues, that will not necessarily work for everybody since people react differently to stress and anxiety. Therefore, it is more important for people to constantly be aware of their stress levels. In addition, it is vital to seek only correct information and avoid inaccurate information on social media.

Misinformation and sensational coverage can invoke a sense of doom. Therefore, it is ideal to trust information only from reputable and authentic sources, such as the WHO and the CDC, and avoid a heavy reliance on social media, unverified news sources, and message forwarding and re-forwarding, if the intended pun will be permitted.

Self-care is important in the pandemic. It is important to be mindful of health and hygiene to avoid infection and maintain mental hygiene. Seeming trivialities like healthy eating, adequate sleep, and regular exercise ought to assume renewed importance during the pandemic. In addition, remaining connected to loved ones, reaching out to others, spending quality time with family, and lending a helping hand to others in the community, are all ways in which people can ensure the emotional connection crucial for an optimal state of mental hygiene.

Care of the Young

The young are particularly prone to anxiety, mainly by observing parents who seem to be under stress. A calm demeanor will go a long way toward helping children. These are the significant signs of stress in children:

- Excessive worry or sadness
- Substance abuse
- Insomnia
- Unhealthy eating
- Excessive crying or irritability
- Lack of attention and concentration
- Poor academic performance
- Bedwetting
- Unexplained headaches

The Healthcare Worker

The average healthcare worker is under tremendous physical and mental stress. Frequent night shifts, protracted absence from family, infrequent meals, exposure to death on an hourly basis, worries about becoming infected with COVID-19, and fears of transmitting it to loved ones, are all taking a great toll on the mental health of healthcare workers. Healthcare workers need to acknowledge that they are under tremendous stress. Experienced senior colleagues are often the best people to help with work-related stress. It is also important to devote time to personal care and family members.

The Disruption of the Social Milieu

The COVID-19 pandemic has affected everyone's life in one way or the other. As a result, a significant swath of the world's population came under lockdown, quarantine, or home isolation. In several countries, offices, schools, and colleges were closed, industries were shut down, and borders were sealed. These unusual measures have resulted in widespread disruptions to life and liberty, ultimately changing the entire social fabric of communities, with attendant job losses pushing millions into hunger and abject poverty.

The appearance of COVID-19 coincided with the Chinese New Year holiday, one of the most celebrated times of the year in China. During this festival, most people migrate to their hometowns. To isolate the people and stop the spread of the pandemic, the Chinese government initially extended the holiday. Simultaneously, to avoid human crowding, Hong Kong Disneyland and Shanghai Disneyland were closed to patrons, like many such public places, which marked the first major disruption to social life. It was closely followed by many such instances worldwide.

The closure of world-famous tourist destinations like the Statue of Liberty, Taj Mahal, the Colosseum, the Eiffel Tower, the statue of Christ the Redeemer, and Ellis Island, and the sealing of borders and travel bans all significantly disrupted the travel and tourism industry. Restaurants, cinema halls, shopping malls, pubs, and markets were closed worldwide. Furthermore, the closure of schools and colleges worldwide disrupted academic calendars. Examinations and graduate-level entry examinations were postponed, with students waiting to see how and when they would be rescheduled and conducted.

The Cancelation of Major World Events

The list of events of global significance that were postponed or canceled sufficiently indicates a world in unprecedented flux. They included the London Book Fair, the Geneva Motor Show, the American College of Cardiology Scientific Session, the World Congress of Cardiology, the Facebook developer conference, etc.

Many major sporting events were either canceled or postponed to safeguard the life of spectators and athletes. The greatest casualty was the 2020 Olympic Games in Tokyo. The IOC president and the prime minister of Japan concluded that the Games of the XXXII Olympiad in Tokyo had to be rescheduled to a date in the summer of 2021 to safeguard the health of the athletes, everybody involved in the Olympic Games, and the international community. Therefore, it was rescheduled from July 23 to August 8, 2021.

Similarly, all professional football tournaments in England, including the Premier League, were halted. Fear and concern for the safety of players, spectators, and organizers led to the suspension of all five of the top leagues in Europe. The UEFA Champions League and Europa League were also suspended. In addition, most international cricket tournaments were canceled or postponed.

The Impact on the World's Major Religions

COVID-19 has had a profound effect on the way religion is practiced all over the world. There have been cancelations of worship services in the various faiths, including the closure of temples, churches, mosques, and synagogues and the cancelation of religious pilgrimages. In addition, the spread of the coronavirus has been linked to certain church congregations in South Korea and the Tablighi Jamaat in Southeast Asia and India.

All places of worship, from China to the Vatican City and from Iran to India, were closed to prevent the virus from spreading further. The authorities in Saudi Arabia urged Muslims to postpone the Hajj until the coronavirus pandemic has been brought under sufficient control. In February, the Kingdom of Saudi Arabia decided to close off the holy cities of Mecca and Medina to foreigners. The Navaratri festival of the Hindus was celebrated more quietly. The temples were largely empty, as devotees decided to celebrate the festival at home.

Churches in many countries were closed, and catholic Masses were postponed. The traditional Mass to celebrate Easter at St. Peter's Basilica in Rome, followed by the pope's Urbi et Orbi blessing in St. Peter's Square, is usually attended by tens of thousands worldwide. Instead, the square was closed off by Italian police, and public Masses and funerals were canceled. Pope Francis live-streamed Easter Sunday Mass from an empty St. Peter's Basilica. He urged the European Union to show solidarity amid the coronavirus pandemic.

Bishops' conferences and dioceses worldwide suspended public Masses. For example, the Diocese of London announced that all Masses and services, including missions and devotions, would be suspended until April 30, 2020, to limit the spread of COVID-19. In addition, the diocese suspended funeral Masses and luncheons. The Catholic Church in France suspended all public Sunday Masses, weddings, and baptisms. Spain, the Netherlands, Belgium, Germany, Slovakia, Malta, and Austria also suspended public Masses.

In the United States, from Seattle to Washington, DC., bishops encouraged the faithful to stay at home, especially those over 60 years old and those with underlying health conditions. As a result, the Archdioceses of Seattle, Little Rock, Salt Lake City, and Santa Fe canceled public Masses. It is, however, certain that regardless of the eventual outcome of the current pandemic, the way religion is practiced worldwide will no longer be exactly the same.

Impact of COVID-19 on Common People

COVID-19 has affected the poor and migrant populations of the world. Major modes of transport, like trains, buses, and airports, suffered closures. With cities under strict lockdown, basic survival became seriously impaired.

In India, millions of migrant workers had to leave the major cosmopolitan centers to return to villages. Many were forced to walk hundreds of miles without knowing where their next meal would come from. Social events like marriages have been postponed. Several families have been separated, with people rendered helplessly unable to look after sick relatives.

Technology has connected the global community like never before. Working from home is now the new normal. There has been a radical shift in work practices, with many nonessential jobs witnessing a move to remote work overnight. Even students attended and, in some places, still attended classes online. The smartphone has increasingly become the window to the world. Now more than ever, it is clear that the digital world will completely take over the analog world.

Impact of COVID-19 on The Global Economy

At the best of times, even when the health impact of an outbreak remains limited, its economic impact can still be huge. As an instructive example, the GDP growth of Liberia witnessed a decline of eight percentage points from 2013 to 2014 during the Ebola outbreak in West Africa, while the country's overall death rate during the same period decreased. A country's economy can bear the brunt of a pandemic in multiple ways. To begin with, a health crisis necessitates a substantial increase in public expenditure to ramp up the healthcare infrastructure needed to treat the infected and control the outbreak.

For instance, special isolation wards are created in public hospitals; free treatment is offered to COVID-19-affected patients, with governments bearing the cost. In special cases, health insurance packages are offered to healthcare workers, as is the case in India, where the government announced insurance coverage of 5 million rupees per person for three months to frontline healthcare workers at the forefront of the COVID-19 battle, facing the highest risk of contracting the disease.

Naturally, rising cases of coronavirus infection also increase the cost of health insurers. By the estimates of Standard & Poor Global, an international credit rating organization, a severe coronavirus pandemic in the US might eventually cost the country's health insurers about US$90 billion in medical claims. Also, preventive measures taken across countries to contain the spread of a pandemic severely impact economic activity, which comes to a grinding halt during an outbreak. Social-distancing efforts lead to the closure of schools, offices, commercial establishments, factories, and transportation and public services. All these grossly disrupt economic and social activities.

In the case of COVID-19, which started in China, the immediate impact was felt in the global supply chain, which faced major disruption, as many factories in the Hubei province, the epicenter of the coronavirus outbreak in China, were shut down, resulting in an immediate decline in China's industrial production and a subsequent delay in the supply of goods from China to other parts of the world.

Most factories in the affected areas remained closed for the better part of January and February 2020. As a result, China's industrial output massively declined by a sharp 13.5% in January and February 2020, compared to the same period in the previous year, with the weakest growth rate recorded since January 1990. Moreover, industrial production continued to decrease in March 2020, with output registering a decline of 1.1%.

Consequently, for the first quarter of 2020, China's overall industrial production declined by 8.4%. In addition, China's exports also contracted during the first three months of 2020, with overseas shipments falling by 17.2% in January and February and 6.6% in March compared to the same months a year earlier.

China is a major global supplier of products and components to leading global manufacturing companies, with nearly 20% of global trade in the manufacturing of intermediate products originating in China, according to the UN's trade and development agency—the United Nations Conference for Trade and Development (UNCTAD). A slowdown in manufacturing activity in China and a decline in exports directly affected industrial activity in many other countries that source critical inputs from China. Sectors with global supply chain linkages such as electronics, automobiles, and chemical products faced imminent raw material shortages and severely hampered production schedules.

Moreover, with the coronavirus gradually spreading to most other countries, including major European and Asian countries and the United States, the manufacturing activities in these countries also started suffering because of similar factory closures or poor work attendance by factory workers. In fact, economic activities nearly halted in many countries that observed a complete lockdown.

The slowdown in manufacturing activity also affected the growth of international trade, as countries faced enormous difficulty exporting or importing goods across borders. For example, UNCTAD estimated that a 2% reduction in exports from China would bring down exports across global value chains by US$50 billion in 2020.

To further compound matters, there has probably been a 5%–15% decline in global Foreign Direct Investment (FDI) flows from previous levels. The countries most severely affected by the pandemic have seen a relatively more adverse impact on their FDI flows than other countries. For example, there was a drop in FDI flows into China, decreasing by 10.8% in January–March 2020, compared to the same period the previous year.

Also, demand for many goods and services has declined sharply during the coronavirus outbreak. For instance, the closure of international borders to contain the spread of the pandemic has directly affected demand in some sectors, such as travel, tourism, and hospitality. For example, according to the International Air Transport Association, airlines stood to lose, in passenger revenues, up to US$113 billion. In comparison, the global revenue from the travel and tourism industry was estimated to decrease by around 17% in 2020 to US$568.6 billion, according to the Mobility Market Outlook on COVID-19, compared to 2019.

The supply of essential and discretionary items has been severely affected by the closure of marketplaces, shopping complexes, and movie halls, which either have been closed in some countries or have suffered reduced attendance. This scenario has badly affected the retail sector and may lead to large-scale layoffs. Entertainment and sports are the other sectors experiencing a similar fallout. In addition, the coronavirus outbreak has impacted the world of live events, with multiple sports and promotional events being canceled or postponed.

Event organizers have incurred huge losses due to the cancellation of events. To make matters even worse, dampened investor confidence levels due to greater uncertainty about the future course of COVID-19 have made the financial markets extremely volatile worldwide, leading to huge market crashes and wealth erosion. This, in turn, has adversely impacted consumption levels. As a result, stock markets worldwide reportedly lost about US$5 trillion in wealth during the first week of March alone, as measured by the MSCI all-country index.

How does one begin to assess the impact of COVID-19 on the world economy? That would prove extremely difficult, given the interplay of so many factors. However, it is undeniable that the magnitude of economic losses due to the pandemic is huge. There is unanimous agreement among leading world economists that the COVID-19 pandemic is having a severe adverse impact on global economic growth. As a result of China being the second largest economy in the world, contributing about 19% to the world GDP, and as a major trade partner of many countries, the slowdown in China's economy is expected to harm the growth of the global economy as a whole.

Policy think tanks and global financial institutions have attempted to estimate the likely economic impact of the novel coronavirus on global economic growth. Many such analysts have attempted to draw analogies with similar health outbreaks from the past, like SARS in 2003, which resulted in a decline of 0.5% to 1% in China's growth in that year, with the global economy suffering a loss of about US$40 billion (or 0.1% of global GDP).

The Organization for Economic Co-operation and Development (OECD) believes that since the global economy today has become significantly more interconnected than it was in 2003 and China has assumed a greater role in global output, trade, tourism, and commodity markets, the economic impact of China's slowdown on other economies will be higher with this pandemic. Moreover, besides the slowdown of China's economy, the growth of individual countries has also been severely impacted due to the slowdown in economic activity in these countries.

The coronavirus will also have a greater impact this time because the global economy was already going through a slow phase before the COVID-19 outbreak. For instance, China grew at 6%—its lowest growth rate since 1990. Therefore, a crisis of this magnitude can translate into a bigger slump for the world economy.

UNCTAD had warned in its Trade and Development Report update, released in March 2020, that many countries in the world would experience a recession. The agency had anticipated a slowdown in the global economy to under 2% in 2020, leading to a global income decline of about US$1 trillion. This cost could go up to US$2 trillion in the likely event that the world economy grows by a meager 0.5%. Similarly, the OECD, comprising 36 countries, had projected that an intensive and long-lasting coronavirus outbreak spreading widely throughout Asia, Europe, and North America could bring global growth down to 1.5% in 2020.

The International Monetary Fund (IMF), in its world economic outlook released in April 2020, drew a completely gloomy picture of the world economy for the year. The IMF projected that the global economy would contract sharply by 3% in 2020 due to the pandemic. The current crisis was much worse than the 2008–2009 financial crisis. The US economy was projected to shrink by 5.9% and the euro area by 7.5%; meanwhile, India and China would grow at a much lower rate of 1.9% and 1.2%, respectively.

The latest OECD estimates suggested that unprecedented measures like the lockdown of cities directly result in a sharp reduction in output, household spending, corporate investment, and international trade. The organization also estimated that there could be a loss of two percentage points in annual GDP growth for each month of strict lockdown or containment measures. If the shutdown were extended for three months, with no offsetting factors, the GDP growth could come down by four to six percentage points from the level that could have been attained under normal circumstances.

Impact on the Job Market

The lockdown observed in various countries led to the shutdown of factories and offices. This caused severe economic disruption, with businesses across several sectors taking significant revenue losses. Naturally, this made people working within these sectors vulnerable to income and job losses.

The International Labour Organization (ILO) reported that nearly 2.7 billion workers, representing about 81% of the world's total workforce, have been affected as a result of the economic fallout of the COVID-19 pandemic; large-scale employment contraction took place in many countries, and the decline in employment is affecting the number of jobs and the aggregate hours of work available. In addition, people working in the most-affected sectors, like retail trade, travel and tourism, and manufacturing, are at a higher risk of job loss.

ILO estimates that about 1.25 billion workers, representing almost 38% of the global workforce, are employed in these worst-hit sectors. The situation is also bad for nearly two billion people working in the informal sector. The people working informally, mainly in developing countries, do not have access to the basic protections of formal jobs, making them the most vulnerable to layoffs.

Impact on the Poverty Index

The COVID-19 pandemic has probably pushed 40–60 million people into extreme poverty. This estimate was made based on the analysis of household survey data and growth projections of 166 countries. By IMF's projections, emerging and developing economies will contract only by 1%, and advanced economies will contract by 6%. However, the research points out that more people will experience the highest increase in extreme poverty since they live close to the international poverty line in developing and emerging countries.

Nearly 23 million people will be pushed to poverty in sub-Saharan Africa and 16 million in South Asia. Three countries are estimated to see the greatest change in the number of the poor—India (12 million), Nigeria (5 million), and the Democratic Republic of Congo (2 million). In addition, more than 1 million people in countries such as Indonesia, South Africa, and China will also be pushed into extreme poverty due to COVID-19.

Impact on Food Security

The transportation and processing of food and food products have been severely affected due to restrictions imposed on the movement of people and goods as a precautionary measure to contain the COVID-19 pandemic. As a result, the UN Food Relief Agency issued a stark warning because the restrictions have made it extremely difficult for people to receive basic food items.

As the world grapples with the coronavirus pandemic, it is also *on the brink of a hunger pandemic*, which could lead to *multiple famines of biblical proportions* within a few months if immediate action is not taken. The UN World Food Program analyzed that about 130 million people could face starvation-like situations by the end of 2020, and this would be an addition to nearly 821 million people who go to bed hungry every night all over the world, and another 135 million people facing crisis levels of hunger due to other reasons.

The Critical Need for Interventionist Measures

The COVID-19 pandemic is not just a major global health crisis. It is also a global economic crisis. Therefore, while it is important to take all possible measures to strengthen healthcare facilities and contain the pandemic, it is equally important to emphasize minimizing the economic impact of the outbreak and putting economies back on track.

Given the adverse consequences of the pandemic on the economic health of countries and the overall well-being of people, policymakers need to devise and implement appropriate measures to provide adequate support to ailing businesses while stimulating demand and job creation in the economy. In addition, specific targeted fiscal and monetary packages also need to be devised for the worst-affected sectors and small businesses, which is essential to save economies from the worst effects of the COVID-19 pandemic.

Countries' severe economic slowdown prompted many to roll out special stimulus packages to revive their economies. In addition, such countries have attempted to extend support directly by increasing government expenditure by way of providing funds to small firms, enabling them to sustain their businesses, keeping people employed, giving specific fiscal packages to the worst-hit sectors, such as aviation and tourism, and increasing fund allocation for infrastructure development.

Some governments have announced the cancellation of certain taxes. Many governments have also deferred payments of term loans, interests, taxes, and utility bills, which has helped to mitigate the immediate financial burden of borrowers and taxpayers. Direct liquidity has also been injected into the system through specific credit lines to small businesses and financial institutions. Such measures have helped by improving firms' cash flow or liquidity position.

Among all the countries that announced fiscal packages, the United States announced the largest package in financial terms, worth US$2.7 trillion, amounting to nearly 13% of its GDP. However, this was not the highest as a proportion of GDP. Japan spent about 21% of its GDP, worth US$1.1 trillion, to help people and businesses survive the COVID-19 crisis. Governments across the European Union announced and implemented large-scale fiscal measures to control the economic consequences of COVID-19. Spain and Italy, the two most-affected countries in Europe, announced stimulus packages estimated to be around 7.3% and 5.7% of their GDP, respectively.

Recognizing the COVID-19 pandemic as an economic emergency, the UK government announced a fresh £330 billion (US$398 billion) of government-backed loans to help businesses impacted by the coronavirus outbreak. As part of the package, smaller businesses could access cash grants of up to £25,000 to minimize the economic impact of COVID-19, and mortgage providers could provide a three-month mortgage holiday.

Although many countries have taken measures at individual levels to tackle the situation on their turfs, they will have to fight the battle consistently with the cooperation of the world's economies, working together to mitigate the negative impact of the pandemic. This is critical for an early revival of the global economy. It will take time to recover from the enormous impact of the COVID-19 pandemic on society, religion, economies, and mental health. Nevertheless, this 21st-century pandemic should allow the world to rethink priorities and strategies. Although the pandemic will decline in due course, especially with the arrival of vaccines, it will also leave deep scars that will continually remind us of human weakness in the face of adversity.

CHAPTER SEVEN

CONTAINMENT AND CONTROL

Critical awareness is the most significant feature in the mitigation and containment of a pandemic. It connotes the possession of accurate knowledge of the real threat of the pandemic and the available resources to fight it. It also demands general coordination between healthcare workers, public health authorities, the diagnostic protocols, communication systems, policymakers, and government agencies at various levels of governance. Well-articulated policy decisions are required to detect and diagnose the disease by appropriate testing, as outlined by the WHO and CDC, while also identifying areas of current disease transmission and deciding on the proper allocation of resources.

Case detection, contact tracing, and surveillance are essential and initial steps in containing the COVID-19 pandemic. Disease surveillance is the backbone of public health since it provides data to understand the pandemic threat, give early warnings, and help formulate a targeted response. Countries with robust public health systems could manage and control the spread and damage that the COVID-19 pandemic caused. However, the lack of appropriate diagnostic kits, in terms of quality and quantity, was a great hindrance in many countries. Therefore, early detection, followed by isolation of affected patients, played the most crucial role in containing the spread.

The lack of vaccines and an overwhelmed healthcare infrastructure at the initial pandemic stage meant non-pharmaceutical measures served as the main modalities of pandemic containment. In the absence of therapeutic countermeasures, public health interventions played a greater role in disease mitigation and containment. The public health measures employed to tackle the COVID-19 pandemic ranged from simple techniques like disease surveillance and hygiene measures to more-restrictive interventions like social distancing, self-isolation, travel restrictions, quarantine, and lockdowns of cities or entire countries.

The pandemic spread at an alarming rate. It became the responsibility of everyone in the world to take appropriate steps to stop its spread and contain the pandemic. It was a war that needed to be fought together at every level.

Prevention of Infection

The Enemy Within

Nothing is more difficult to fight than an invisible enemy. Unfortunately, the coronavirus is that invisible enemy. The first step to fighting coronavirus is to learn about it and the disease it causes—COVID-19. The second step is to learn how to prevent infection. Unfortunately, there is a deluge of information about the coronavirus in the media. Yet, it is important to differentiate between genuine and false information.

As matters stand, social media has done more harm than good by spreading misinformation and fake news. Relevant articles published in reputable medical journals are beyond the comprehension of a layman who has inadequate education about medical terms to grasp the technical language used. So, the most reliable sources of information are those published by the WHO and the CDC, which are readily available on their respective websites. Moreover, they are written in plain language so everybody can understand them.

Information about how infection can be prevented from spreading from one person to another assumes the next level of importance. Droplet spread and direct contact with a COVID-19 patient transmit the new coronavirus; therefore, stringent hygienic measures are employed to prevent respiratory infections and were applied to the Spanish flu and SARS. These include using face masks and gloves, hand washing, and proper etiquette for coughing and sneezing. However, educating the general public on how to use these measures properly is important.

Hand Washing

Wash your hands frequently, using soap and running water for 20 seconds. If your hands are not visibly dirty, clean them frequently using an alcohol-based hand rub. Do not touch your eyes, nose, and mouth with unclean hands. Clean your hands properly, especially in the following situations:

- After sneezing or coughing.
- After coming home from outside or meeting other people, especially if they are ill.
- After caring for the sick.
- Before preparing food, eating, or feeding children.
- Before and after using the bathroom.
- When hands are visibly dirty.
- Before and after changing diapers.
- After handling animals or animal waste (pets also develop coronavirus infections).

COVID-19 can survive for two to three days on plastic and steel surfaces. Therefore, there is always a high probability of touching contaminated surfaces and promptly getting infected. A quick scrub and a rinse are insufficient to eliminate all virus traces on your hands. Below is a step-by-step process for effective hand washing:

Step 1: Wet hands with running water.

Step 2: Apply enough soap to cover wet hands.

Step 3: Scrub all surfaces of the hands, including the back of hands, between fingers, and under nails, for at least 20 to 30 seconds.

Step 4: Rinse thoroughly with running water.

Step 5: Dry hands with a clean cloth or single-use towel.

Hot water is not required for washing hands. However, your hand sanitizer should contain at least 60% alcohol; you will need to rub it into your hands for at least 20 seconds. CDC recommends using alcohol-based hand sanitizer with more than 60% ethanol or 70% isopropanol in healthcare settings. Also, use paper towels or clean clothes to dry your hands completely.

Face Masks

Face masks are effective at preventing coronavirus infection. Since the virus spreads by droplets from the mouth and nose, masks, which cover these areas, are extremely beneficial when you are near a person with a cough and fever. However, face masks were always in short supply worldwide, and local authorities shared guidelines for their use.

In addition, you should understand that you are not fully protected only by wearing masks. It is important to stress this point as *masks tend to induce a false sense of security*. Similarly, using masks incorrectly will neither provide adequate protection nor prevent spreading the virus to others. If you use masks, you should follow best practices about how to wear, remove, and dispose of them and hand hygiene after removal. Medical masks are mandatory for those with cough, fever, and breathlessness; healthy people looking after a person suspected to have contracted SARS-CoV-2 must wear them. It is also mandatory for healthcare workers who come into contact with patients outdoors, indoors, and in ICUs.

Mask Management

Mask management is much more important than generality often realizes. They are helpful if used properly. However, an improperly used mask may be a potent source of infection. This is how to use a face mask correctly:

- Before wearing a mask, wash your hands properly.
- Place the mask carefully, covering your nose and mouth without leaving any gap between the mask and your face.
- Tie the mask tightly.
- Do not touch the mask with your hands.
- While removing it, do not hold the front part of the mask.
- Untie it from behind.
- After removing the mask, dispose of it immediately in a closed bin.
- Wash your hands with soap and water after you remove the mask.
- Do not reuse a single-use mask.
- Replace the mask if it becomes damp.

Whatever the guidelines, if there is no shortage of face masks, those leaving their homes should use them, especially in crowded locations.

Household Disinfectant

It is crucial to disinfect household objects frequently. Repeatedly touched objects like door handles, knobs, electric switches, car doors and handles, and tabletops must be cleaned with a disinfectant that is effective against the novel coronavirus. One need not purchase a costly disinfectant. It can be easily prepared at home.

Making A Bleach Solution

Mix as follows: 5 tablespoons (1/3 cup) bleach per gallon of water or four teaspoons bleach per quart of water. The bleach solution effective against coronavirus must contain 0.1% sodium hypochlorite. Also, you must follow the manufacturer's instructions for application and proper ventilation. Before preparing the solution, check the expiration date. Never mix household bleach with ammonia or any other cleanser. Alternatively, a solution containing at least 60% alcohol may be used.

Social Distancing and Self-Isolation

Social distancing and self-isolation are two methods of containing pandemics that have been used widely. The Spanish flu of 1918 provided evidence of the importance of social distancing. Those parts of the US that banned public gatherings and closed theaters, schools, and churches early had far lower peak death rates.

Moreover, social distancing means creating a barrier between two people to prevent the transmission of a contagion. The lessons learned from the Spanish flu have proved very useful in containing the COVID-19 pandemic. Social-distancing rules and norms are imposed strictly by some countries and loosely by others. Similarly, some people strictly follow the rules of social distancing, whereas many others flout the rules. In some countries, legal steps have been taken against such violators.

Social Distancing

In the absence of vaccines, social distancing is probably the only means to break the chain of transmission to contain the pandemic. Consequently, different countries have implemented different norms and regulations for social distancing, including a complete ban on mass gatherings, the closure of public places like malls, parks, gyms, and restaurants, curfews, and total lockdowns of cities and entire countries.

For a densely populated country like India, and in the absence of less-than-salutary health infrastructure, locking down the country entirely went a long way in limiting the number of cases and casualties. On the other hand, failure to ensure or achieve complete lockdown—a feat that is not easy to accomplish—has most likely resulted in huge cases and mortalities recorded in countries like Italy and the US. However, as death tolls mounted and the fear of COVID-19 became the beginning of wisdom, people automatically shunned public gatherings and voluntarily isolated themselves.

Self-Isolation

Self-isolation is social distancing with a subtle difference. It creates a barrier between an infected person and a healthy one. The spread of the virus is limited by isolating the infected person. COVID-19 is extremely contagious, and one infected person is capable of infecting many more. The maximum incubation period of COVID-19 is 14 days. Therefore, a minimum isolation period of 14 days is needed to prevent the spread of infection. A person suspected of being infected should self-isolate at home in a separate, well-ventilated room, dependent on family members.

Flattening The Curve

A major objective of social distancing is to flatten the curve of the epidemic. It means the slow spread of the virus in the community because of social distancing, which will reduce the peak number of cases during the epidemic. A flat curve means fewer people are affected at any given time. On the other hand, an inability to flatten the curve means that too many people are getting infected simultaneously, overwhelming the healthcare infrastructure, as happened in parts of the US, Italy, and China.

The rules of social distancing during the lockdown varied from one country to the next. However, the following rules were in place in almost every country. Everyone should remain at home and could leave their respective homes only for the following reasons:

- For medical needs.
- During emergencies, medical or otherwise.
- To buy essential food items and medicine.
- To care for sick and vulnerable people.

However, those on the frontline of the battle against COVID-19, like doctors and other healthcare workers, sanitation workers, law enforcement personnel, and those in essential services like water supply and sanitation, were allowed to leave home. But, when they left home, they were constrained to keeping a minimum distance of 2 meters from one another.

Shielding

Vulnerable groups are always available and are not encouraged to mix with others during a pandemic. No one should visit them either because they are prone to contracting the coronavirus. The NHS in the UK urged people with serious health issues not to go out for at least 12 weeks. This process is known as shielding. These are people who make up the vulnerable groups:

- Cancer patients.
- Those who are on steroids. Steroids reduce immunity.

- Organ transplant recipients.
- Those with certain genetic diseases.
- Those who have a serious respiratory illness.
- Pregnant women with heart diseases.

Protecting Others from COVID-19

An individual must take care of himself and take all possible precautions to prevent the spread of the novel coronavirus. Also, it is the prime responsibility of every person to protect others. Therefore, an individual must try to comply with the following measures to protect others:

- While coughing or sneezing, cover your nose and mouth with a flexed elbow or a tissue rather than your palms.
- Throw the tissue into a closed bin immediately after use.
- Avoid close contact with others if you have a cough or fever.
- Do not spit in public.
- Seek early medical care if you are sick.
- Do not shake people's hands.

Yet, we must accept that it is not easy for anyone to practice social distancing and self-isolation for a long time. Loneliness, emotional isolation, social and economic disruption, and infringements on civil rights and liberties are all important consequences of social isolation.

Quarantine

Quarantine is one of the oldest methods of controlling epidemics and pandemics. The practice of quarantine began in 14th-century Europe in response to the epidemic, the Black Death. During the epidemic of the Black Death, ships arriving at ports were quarantined to prevent the spread of disease in cities and towns.

Quarantine means restricting the movement and activities of asymptomatic people exposed to a confirmed case of an infectious disease to prevent further spread. Quarantine centers were often established on islands on the continental shelves of countries.

Quarantine is essentially *cordon sanitaire*—a guarded line preventing anyone from leaving an area infected by a disease and thus spreading it. Establishing cordon sanitaire is an essential element in the fight against COVID-19. Thousands of people have faced quarantine, and many more will face it as long as COVID-19 remains with us. Some people are quarantined in their homes, while others are quarantined at a designated facility. Wherever people are kept, they are strictly isolated from a confirmed case of COVID-19.

Quarantine is one of the most complex and controversial public health measures. On one end, it involves significant curtailment of individual civil liberties, while on the other end, it is in the best interests of society. Expectedly, it generates conflict between an individual's rights and the state's enormous power to curtail those rights. Therefore, it can only be justified if the benefit to society far outweighs the harm to individuals. Naturally, a fine and critical balance between the two becomes essential.

In the final analysis, quarantine should not spark tension between citizens and the state. It should be used only when careful consideration by public health specialists recommends its use. Experience is probably the best guide on this. Governments bear the enormous responsibility of implementing quarantine with transparency. The transparency demonstrated by a government will go a long way in ensuring the public's acceptance of quarantine.

Suffice it to say that the best form of quarantine is voluntary quarantine. During quarantine, it becomes the responsibility of the government to provide adequate medical facilities and psychological support, not just items for daily needs. Quarantine disrupts people's lives and liberties and involves a huge expenditure. The government should bear the economic cost of quarantine and disruptions to economic activity, which should not be passed on to the citizens.

Border Control

The only way an epidemic becomes a pandemic is through the movement of people, animals, and goods across international borders. Air travel is particularly incriminated in this process, being one of the most efficient means of spreading infection from one country to another, as observed with the COVID-19 pandemic.

Governments have tried to close borders and impose travel bans to contain the pandemic. Yet, such a massive disruption is not always entirely possible. Public health law, especially the WHO's International Health Regulations, provided a legal framework for governments worldwide regarding travel bans and border closures. The COVID-19 pandemic triggered travel bans and border closures in several countries to contain the pandemic. Some countries took that measure early enough in the pandemic, while some did it late during the pandemic.

A few countries, for instance, India, shut down their borders completely and imposed a total ban on airlines and foreign travel with only a few exceptions. Such transnational measures have included screening at entry and exit, collecting and disseminating passenger information, medical checkups at airports, and quarantining and isolating sick passengers. As a result, passengers have suffered severe disruptions in their travel plans, with many stranded in foreign countries. Despite these drawbacks, governments must implement transnational measures to halt the march of the pandemic within the limits of the International

Health Regulations set by the WHO.

In conclusion, the efficacy of preventive and containment measures will always depend on compliance. Many challenges have come up, including shortage of masks, gloves, and sanitizers, social and psychological disruptions, economic crises, infringement on civil liberties, etc. Even as early as the start of the pandemic, public health authorities worldwide agreed that until effective vaccines and medicines were discovered, those containment measures remained mankind's only hope for preventing infection and the global spread of the coronavirus. A vaccine was largely accepted as the best solution to the pandemic. Thankfully, the WHO coordinated with various stakeholders in their effort to develop effective vaccines, which later became available.

CHAPTER EIGHT

THERAPEUTIC OPTIONS AND THE SCIENCE OF VACCINES

"Vaccines provide direct protection by reducing susceptibility to disease or infection. Vaccines provide indirect protection by reducing the number of people infected in a population or their infectiousness. These vaccine effects can be assessed in clinical trials by measuring the efficacy to prevent disease, to prevent infection, and to reduce infectiousness, as well as in studies to assess indirect effects of the vaccine."

—M. E. Halloran, I. M. Longini Jr., C. J. Struchiner, Design and Analysis of Vaccine Studies (Springer, 2010).

In simple terms, if it initially seemed as if the COVID-19 pandemic was propelling mankind toward a disaster of destruction, the rapid emergence of vaccines saved the day. The phrase *rapid emergence* is significant. This is because the apparent speed of roll-out of vaccines, at least within the comparative context of what was previously known as the snail-slow speed of the development and authorization of vaccines, largely formed the basis of the politics of controversy and suspicion of COVID-19 vaccines.

Yet, while it may seem as if the COVID-19 vaccine moved rather quickly to emergency use authorization and approval, the reality is that researchers and scientists were able to draw upon years of data, research, and advances to fast-track these lifesaving vaccines. As James Hoffman, PharmD, chief patient safety officer at St. Jude Children's Research Hospital, Memphis, Tennessee, said, *"All science builds. The fact that we were able to make these vaccines available so quickly can be traced back to many different scientific advances."*

Knowledge of Similar Viruses

In several ways, researchers were already at a significant vantage point in the race to create effective vaccines for COVID-19. SARS-CoV-2 is the novel virus that causes Coronavirus disease 2019 (COVID-19). That means it wasn't known or recorded before it emerged; however, that doesn't mean the world of virology had never encountered a virus like it before 2019.

An article published in *Nature* explains that SARS-CoV-2 is no longer fundamentally different from other viruses that have appeared in humans in the past, but none of them has spread like SARS-CoV-2. SARS-CoV, more commonly known as Severe Acute Respiratory Syndrome (SARS), appeared between 2002 and 2004. MERS-CoV (Middle Eastern Respiratory Syndrome or MERS) first appeared in 2012. These viruses are important because they bear many similarities to SARS-CoV-2. The vital fact is that researchers were already working on vaccines to treat them.

According to *Nature*, two vaccines for SARS were never offered to the public because they were still in Phase I clinical trials. Therefore, because of existing and subsisting research about how to optimize vaccination against SARS and MERS, researchers working on COVID-19 vaccines already knew that targeting the spike protein would be an effective way to help people make antibodies to protect against COVID-19. That basic conceptual strategy helped fast-track the development and roll-out of the COVID-19 vaccines.

Prevention Strategies and Therapeutic Options Against COVID-19

The treatments that have been in use are targeted to achieve lowered SARS-CoV-2 burden in patients, alleviate severe symptoms, and slow the progress of COVID-19. The spectrum of medical therapies to treat COVID-19 is growing and evolving rapidly, including drugs approved by the US Food and Drug Administration (FDA) and drugs made available under the FDA Emergency Use Authorization (EUA). Therefore, the Centers for Disease Control (CDC) has strongly encouraged clinicians, patients and their advocates, and health system administrators to regularly consult the guidelines published by the National Institutes of Health (NIH). These guidelines' treatment and management recommendations are based on scientific evidence and expert opinion and are frequently updated.

The current clinical management of COVID-19 consists of *infection prevention, control measures,* and *supportive care*, including supplemental oxygen and mechanical ventilatory support when indicated. FDA has approved one drug, remdesivir (Veklury), for treating COVID-19 in certain situations. It has been firmly established that effective early treatment of COVID-19 can help avert progression to more serious illness, especially for patients at high risk of disease progression and severe illness, with the additional benefit of reducing the burden on healthcare systems.

Several novel therapeutics, for instance, monoclonal antibodies, are available under Emergency Use Authorization (EUA) for early outpatient treatment. Trials to assess the potential effectiveness of these therapeutics in outpatients at high risk of disease progression are currently ongoing. Summarily, the novel therapeutics that are currently available include convalescent plasma, monoclonal antibodies, and repurposed drugs that are already approved. For example, Veklury (remdesivir) was the first drug approved by the FDA to treat COVID-19, followed by Emergency Use Authorization (EUA) of monoclonal antibodies, bamlanivimab, and casirimab/imdevimab.

The antiviral and chemotherapeutic agents used in the current clinical trials directly act on COVID-19, limiting the viral replication in host cells. The immune system is also boosted using cell-based therapy, convalescent plasma treatment, or inhibition of inflammatory processes linked with cytokine storms. Other antivirals currently under clinical trials include Favipiravir, Lopinavir, Umifenovir, Camostat, Hydroxychloroquine, and Azithromycin. The efficacy of anti-inflammatory drugs such as glucocorticoids and Thalidomide is also currently under assessment.

Vaccines And Vaccination

Types of Vaccines That Were Developed for SARS-CoV-2

1. Inactivated vaccine that uses the native virus but is rendered replication-deficient through heat or chemical treatment.

2. Live-attenuated vaccine in which the virus can replicate, but in such a limited manner that it cannot precipitate the disease condition.

3. Subunit vaccine that incorporates only subsections of the native virus, such as the S protein.

4. Viral vector vaccine that encapsulates the genome of a different, weakly pathogenic virus, with additional DNA that encodes the target viral antigen.

5. DNA vaccine using a DNA plasmid that encodes the target antigen, and this is often administered by electroporation.

6. RNA vaccine of RNA, encapsulated within an LNP to decrease RNA degradation and increase translation efficiency.

Vaccine: The Historical Perspective

Over a century ago, vaccines significantly increased life expectancy, fundamentally reshaping communities and their economies all over the globe. As vaccinations have become widely available and used, the wholesome result is that the devastating impacts of many infectious diseases have faded. However, treating infectious diseases can be astronomical in cost, as demonstrated by the seasonal flu. This huge yearly economic and social burden has to be borne if thousands of lives are to be saved.

From all empirical indicators over the past ten decades, widespread preventative vaccination, apart from serving to help lower such costs, can also play a pivotal role in protecting people from viral infections efficiently and sustainably, resulting in the complete elimination or significantly reduced transmission within the herd population.

Vaccines and Severe Acute Respiratory Syndrome Coronavirus 2 (SARS-CoV-2)

SARS-CoV-2 has afflicted millions of people worldwide. As of January 18, 2021, on the global scale, 96 million cases had been seen, and 2 million deaths had been recorded. In a nutshell, these daunting statistics dictated that a safe and effective vaccine was urgently needed if the human devastation was to be stemmed.

The Food and Drug Administration (FDA) approved two investigational vaccines sponsored by Pfizer and Moderna. It is pertinent to note that both Moderna and the BioNTech/Pfizer partnership employed similar techniques for the conceptualization, development, and production of their vaccines by utilizing mRNA that encodes for subunits of the SARS-CoV-2 S protein. Therefore, the FDA approved two Covid vaccines for Emergency Use Authorization (EUA)□the Pfizer vaccine and the Moderna vaccine. But first, let us take a crash course on the genetic structure of SARS-CoV-2.

Genetic Sequence of SAR-CoV-2

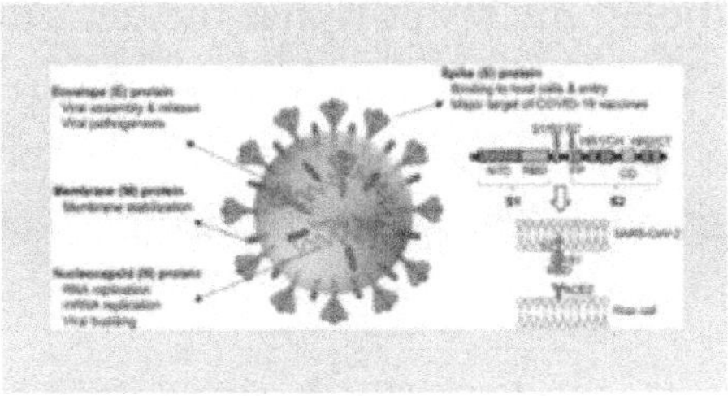

SARS-CoV-2: An enveloped single-stranded RNA (ssRNA) virus; spike-like-glycoproteins expressed on the surface in a *corona* formation.

The SARS-CoV-2 genome: 79.6% genetic identity to **SARS-CoV**. It consists of **four key proteins:**

1. **The S protein (S)** enables the attachment and entry of SARS-CoV-2 to the host cells.

2. **The membrane (M)** protein is an integrity component of the viral membrane.

3. **The nucleocapsid (N)** protein binds to the viral RNA and supports the nucleocapsid formation, assisting in virus budding, RNA replication, and mRNA replication.

4. **The envelope (E)** protein is the least understood for its mechanism of action and structure but seemingly plays roles in viral assembly, release, and pathogenesis.

Structure of SARS-CoV-2

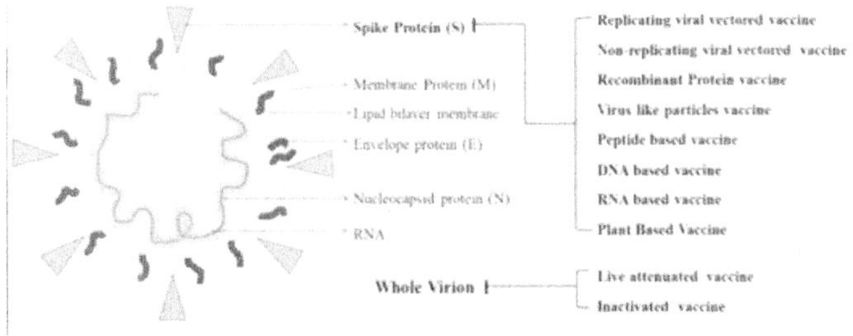

IMPORTANT: Its key structural protein is a target for vaccine

The BNT162b2 mRNA COVID-19 vaccine (The Pfizer Vaccine)

BNT162b2 mRNA Vaccine is a lipid nanoparticle formulated, nucleoside-modified RNA (modRNA) that encodes membrane-anchored SARS-CoV-2. It is a full-length spike protein, modified by two proline mutations to lock it in the perfusion conformation. Over 43,548 participants were randomized in the clinical trial study for the celebrated Pfizer vaccine study, with 21,720 receiving the BNT162b2 and 21,728 receiving a placebo. It was conclusively found that a two-dose regimen of BNT162b2 was 95% effective against COVID-19 in people of ages 16 years and older.

In practical terms, the vaccine is 90-100% effective. The two 30 μg doses of BNT162b2 elicited high SARS-CoV-2 neutralizing antibody titers, robust antigen-specific CD8+, and Th1-type CD4+ cells response. Local and systemic reactions include mild to moderate pain at the injection site and pain (much less in older people > 55 years); sometimes, more severe symptoms are reported in younger recipients, including fatigue, headache, and fever > 38 degrees celsius. More adverse events were reported in 0.1% of cases, and lymphadenopathy in 0.3%. Others had varying degrees of shoulder injury and numbness.

The Moderna vaccine (mRNA vaccine)

The Moderna mRNA vaccine-1273 is a lipid-nanoparticle (LNP)-encapsulated mRNA vaccine expressing the prefusion-stabilized spike glycoprotein. In late July 2020, a clinical trial assessed the safety and efficacy of the mRNA-1273 vaccine in preventing SAR-CoV-2 infection, culminating in Phase 3 randomized and stratified placebo-controlled trial. Vaccine efficacy was found to be 94.5 %.

The vaccine provides a sterile liquid at a concentration of 0.2mg per milliliter, and second injection, 28 days later, of 0.5ml containing 100microgram of mRNA 1273. Storage is at 2-8 degrees (35.6 to 46.4 F). Requiring no dilution, the dose can be held in a syringe for 8 hours at a stretch.

How mRNA COVID-19 Vaccines Work

Given the extreme suspicion that still surrounds the roll-out of vaccines, this attempt to explain how vaccines work is rendered in simple language to allow for layman comprehension. To trigger an immune response, many vaccines we have known from the past introduce a weakened or inactivated germ into our bodies.

However, this is not what mRNA vaccines do. Instead, mRNA vaccines use mRNA created and developed in a laboratory to teach our cells how to make a protein, or just a piece of a protein, that triggers an immune response inside our bodies. That immune response, which produces antibodies, is what helps to protect us from getting sick from that germ in the future.

1. First, mRNA COVID-19 vaccines are given in the upper arm muscle. After vaccination, the mRNA will enter the muscle cells. Once inside, they use the cells' machinery to produce a harmless piece of the spike protein. The spike protein is found on the surface of the virus that causes COVID-19. After the protein piece is made, our cells break down the mRNA and remove it.

2. Next, our cells display the spike protein piece on their surface. Finally, our immune system recognizes that the protein does not belong there, thereby triggering the immune system to produce antibodies and activate other immune cells to fight off what it thinks is an infection.

3. At the end of the process, our bodies have learned how to help protect against future infection with the virus that causes COVID-19. The benefit is that people get this protection from a vaccine without risking the serious consequences of getting sick with COVID-19. In addition, any side effects from receiving the vaccine are normal signs that the body is building protection.

These are the facts about mRNA COVID-19 vaccines:

- mRNA COVID-19 vaccines cannot give us COVID-19 or other illnesses.
- mRNA vaccines do not use any live virus.
- mRNA vaccines cannot cause infection with the virus that causes COVID-19 or other viruses.
- mRNA vaccines do not affect or interact with our DNA in any way.
- mRNA from these vaccines does not enter the cell's nucleus where our DNA (genetic material) is located, so it cannot change or influence our genes.
- The mRNA and the spike protein do not last long in the body.
- Our cells break down mRNA from these vaccines and get rid of it within a few days after vaccination.
- Scientists estimated that the spike protein, like other proteins our bodies create, may stay in the body for up to a few weeks.
- mRNA COVID-19 vaccines have been rigorously evaluated for safety and effectiveness.
- The only COVID-19 vaccines the Food and Drug Administration (FDA) makes available for use (by approval or Emergency Use Authorization) are those that meet these standards.
- While COVID-19 vaccines were developed rapidly, all steps and measures have been taken to ensure their safety and effectiveness.

mRNA Vaccines Are Newly Available to the Public but Have Been Studied for Decades

Researchers have been studying and working with mRNA vaccines for decades. Interest has grown in these vaccines because they can be developed in a laboratory using readily available materials. This means vaccines can be developed and produced in large quantities faster than with other methods for making vaccines. mRNA vaccines have been studied before for flu, Zika, rabies, and cytomegalovirus (CMV).

As soon as the necessary information about the virus that causes COVID-19 was available, scientists began designing the mRNA instructions for cells to build the unique spike protein into an mRNA vaccine. Future mRNA vaccine technology may allow one vaccine to provide protection against multiple diseases, thus decreasing the number of vaccine shots or administration needed to protect against common vaccine-preventable diseases. Beyond vaccines, cancer research has used mRNA to trigger the immune system to target specific cancer cells.

BioNTech/Pfizer and Moderna: A Comparative Analysis

BioNTech/Pfizer and Moderna encapsulate their RNA vaccines within LNPs, which may enable cytoplasmic delivery *via* fusogenic mechanisms. Additionally, cytoplasmic delivery may improve translation efficiency, yet it may also decrease RNA immunostimulation. Since RNA stimulates the immune system, it acts as an adjuvant by activating specific toll-like receptors (TLRs), mainly TLRs 3, 7, and 8, located within the cell's endosomes.

The differences in major symptoms are that, while the Moderna trial revealed pain, headache, and chills, the BioNTech/Pfizer vaccine mainly caused pain, fatigue, and headache. In addition, Moderna tested antibody response through ELISA assays regarding functional assessment, while BioNTech/Pfizer utilized an RBDbinding IgG assay. They both require booster injections, and while Moderna will continue to boost on day 29 after the initial injection, BioNTech/Pfizer will boost on day 21 after the initial injection.

Oxford/AstraZeneca and CanSino: The Non-replicating Viral Vector Vaccines

The Oxford/AstraZeneca, AZD1222 employed a different viral vector—an Aenovirus derived from the chimpanzee and was subsequently named AZD1222. Compared with Moderna and Pfizer vaccines, Ad5-nCoV specifically encodes for the full-length S protein of SARS-CoV-2. Both Moderna and BioNTech/Pfizer encode subunits of the S protein.

The Rapid Development of the Vaccines

The rapid development of vaccines in response to a raging pandemic has generated a lot of controversies fueled by various conspiracy theories, ranging from the ridiculous to the sublime. Yet, the fact must be noted that the utilizable science was already established and ready for application. In other words, even though these are new vaccines, their production methods had been developed decades ago.

Without a doubt, tremendous resources were deployed to this global effort, including time, money, and supplies. The studies were enormous: 44,000 participants for Pfizer and 30,000 for Moderna. More than 37% of the Moderna participants were from minority groups. COVID-19 had exploded in the number of infections very quickly. As soon as the test patients were vaccinated, they were quickly exposed to such a large inoculation dose of disease that it was easy to confirm that the vaccines were effective.

A Thorough Approval Process

Emergency use authorization for COVID-19 vaccines progressed rapidly, largely due to a perfectly understood need to quickly produce an effective vaccine against the novel coronavirus. Communication, collaboration, and resources were fundamental to this herculean effort. The combined efforts of many researchers worldwide, working from existing technologies, allowed the vaccines to be developed and tested in a shorter timeframe.

Yet, the speed did not mean researchers and scientists were not putting the population's safety on the front burner during the process. Getting a vaccine authorized was still a rigorous process. In fact, at one point, more than 180 COVID-19 vaccines were in development. But in the United States, only three vaccines are currently authorized by the FDA.

In late August 2021, the Pfizer/BioNTech vaccine received full FDA approval for use in people of ages 16 and older. Still, in the words of James Hoffman, PharmD, *"From a process standpoint, the failure of vaccine candidates illustrates the rigor that was involved in developing, testing, and authorizing these vaccines."* Yet, while all data points to COVID-19 vaccines as being safe and effective, work on monitoring the vaccines continues. The Food and Drug Administration (FDA), CDC, researchers, and developers are continuing to gather information about the safety and efficacy of COVID-19 vaccines. According to Dr. Hoffman, *"The science is not stopping. There are different layers of reviews and analysis happening. We are constantly learning about the vaccines. We continue to learn more about the vaccines every day."* One fact is undeniable—the benefit of COVID-19 vaccines outweighs the risks. Data backs up this unequivocal assertion.

Pfizer Vaccine Safety and Efficacy: Myths and Confusion

Some of such myths and conspiracy theories include: the side effects from the COVID-19 vaccine will be severe, if not much worse than having the virus; many have already died from COVID-19 vaccine trials; if one has already contracted COVID-19 and recovered, one doesn't need the vaccine because one is already immune; the COVID-19 vaccine was developed as a way of controlling the general population, through either microchip tracking or nano-transducers in the brain; the COVID-19 vaccines were developed using fetal tissue; one might be allergic to the vaccine and not know until one is vaccinated.

None of these spurious claims are true. No specific safety concerns were identified in sub-group analyses by age, race, ethnicity, medical comorbidities, or prior SARS-CoV-2 infection. Local site reactions were the most frequent complications, and most were in the mild to moderate range: Injection site pain (84.1%), Fatigue (62.9%), Headache (55.1%), Muscle pain (38.3%), Chills (31.9%), Joint pain (23.6%), Fever (14.2%). There were six deaths in the study period—2 in the group of patients who received the vaccine and 4 in the group of patients who did not receive the vaccine (placebo). Out of 44,000 people, these were equal to the rate of death seen in the general public.

Many believe that the COVID-19 vaccine projects were rushed, so the vaccines will not be safe. Nothing could be further from the truth. What is true is that the COVID-19 vaccine was rapidly developed for several reasons, the most significant being the alarming devastation that it was wreaking on human mortality and the economies of nations.

Two vaccines, Pfizer and Moderna, that were shown to be effective, are based on novel messenger RNA (mRNA) technology. In contrast, the Oxford/Astrazeneca vaccine is based on a recombinant adenovirus vaccine. Researchers have been working with mRNA vaccines for over ten years. Additionally, researchers have learned to develop more rapid ways of conducting clinical trials by working on multiple studies.

Most of the COVID-19 vaccine studies, Phase 1 (first in humans; initial safety) and Phase 2 (safety, dose, and immunogenicity), were intricately designed in one study. If Phase 1 indicates adequate safety, Phase 2 will start immediately without having to design a new study and wait for more regulatory hurdles. The researchers also designed Phase 3 (efficacy) studies while the phase 1 and 2 trials were being conducted, based on empirical assumptions that the vaccines would ultimately be safe and effective.

The most effective way to stem the spread of the coronavirus is through personal behavior—mask-wearing, hand washing, and social distancing. The more Americans can practice social distancing and other precautions to protect against coronavirus transmission, the quicker the US will get the coronavirus under control. Additionally, in the US and other countries, efforts must be robust and swift towards a combination of coronavirus testing, the isolation of people who tested positive for the virus, and contact tracing. The long-term COVID-19 pandemic's future will largely depend on the strength of immunity people build against the virus after exposure and the administration of the vaccines that have been approved.

Although there are many possibilities, ranging from seasonal surges in infection to the development of herd immunity, the clear fact is that there is so much we still do not know about this virus.

CHAPTER NINE

THE MENACE OF VARIANTS

All viruses, including SARS-CoV-2 that causes COVID-19, change over time. Most of such changes have little or no impact on the virus's properties. However, some changes may affect the virus's properties, such as how easily it spreads, the associated disease severity, or the performance of vaccines, therapeutic medicines, diagnostic tools, or other public health and social measures.

The WHO, in collaboration with partners, expert groups, governments, institutions, and researchers, has been monitoring and assessing the evolution of SARS-CoV-2 since January 2020. As a result, variants of the coronavirus started to emerge in late 2020, posing an increased risk to global public health and prompting the characterization of specific *Variants of Interest (VOIs)* and *Variants of Concern (VOCs),* the principal aim being to prioritize global monitoring and research and ultimately to both inform and enhance the ongoing response to the COVID-19 pandemic.

The WHO and its international networks of experts consistently monitor changes to the virus so that if significant amino acid substitutions are identified, countries and the public can be promptly informed about modifications that may be needed in response to the variant and thereby prevent its spread.

Globally, systems have been established and continually strengthened to detect *signals* of potential VOIs or VOCs and assess these based on the risk posed to global public health. Naturally, governments and their local health authorities may choose to designate variants of local interest or concern. However, it is acknowledged that reducing transmission through established and proven disease control methods is vital to the global strategy for reducing the occurrence of mutations that might have negative public health implications.

Current strategies and measures recommended by the WHO continue to mitigate against those virus variants identified since the beginning of the pandemic. Empirical evidence from various countries that have reported extensive transmission of VOCs indicates that Public Health and Social Measures (PHSM), including Infection Prevention and Control (IPC) measures, have effectively reduced COVID-19 cases, hospitalizations, and deaths.

In the circumstances, national and local authorities are encouraged to continue strengthening existing PHSM and IPC measures. It is also vital to continue strengthening surveillance and sequencing capacities while applying a systematic approach to provide a representation of the extent of transmission of SARS-CoV-2 variants based on local context and to detect unusual epidemiological events.

Having said all that, since December 2020, several coronavirus variants have been identified, with each new variant raising pertinent questions, such as:

- Are people exposed to a greater risk of falling sick?

- Will the current arsenal of COVID-19 vaccines be effective against these variants?
- What new or different things should people be doing to remain safe?

The Science of Coronavirus Mutations

Variants of viruses occur when there is a change, or mutation, to the virus's genes. It is the nature of RNA viruses, such as the coronavirus, to evolve and change over time. Invariably, geographic locations result in genetically distinct variants. Therefore, it is important to realize that mutations in viruses, including the coronavirus that causes COVID-19, are neither new nor unexpected. All RNA viruses mutate over time and some more than others. Therefore, flu viruses often change, which is why one is recommended to get a new flu vaccine every year.

Variants of Concern (VOC) and Variants of Interest (VOI)

Coronavirus variants are classified into different categories by the WHO and the Centers for Disease Control and Prevention (CDC). A variant of interest (VOI) is a coronavirus variant that, compared with earlier forms of the virus, has genetic characteristics that predict greater transmissibility, evasion of immunity or diagnostic testing, or more severe disease. A variant of concern is more infectious and more likely to cause breakthrough infections or reinfections in those vaccinated or previously infected. In addition, these variants are more likely to cause severe disease, evade diagnostic tests, or resist antiviral treatment.

Alpha, beta, gamma, delta, and omicron variants of the SARS-CoV-2 coronavirus are classified as variants of concern. A variant of high consequence is a variant for which current vaccines do not offer protection. At least for now, no SARS-CoV-2 variant has been classified as a variant of high consequence.

Vaccines And New Variants

There is evidence from laboratory studies that some immune responses driven by current vaccines could be less effective against some variants. That is why those eligible for COVID-19 booster vaccines should get them for added protection against infection and severe disease. Yet, it must be carefully noted that the immune response involves many components, including B cells that make antibodies and T cells that can react to infected cells. Therefore, reducing one does not necessarily mean that current vaccines will not offer protection.

What is important from the perspective of public health is that those who have been vaccinated should be on the alert for changes in guidance from the CDC and continue with coronavirus safety precautions to reduce the risk of infection, such as mask-wearing, physical distancing, and hand hygiene. Since mutations are dealt with every year for the flu virus, even if there are major mutations of the coronavirus, the vaccine development process is probably versatile enough to accommodate changes.

The Omicron Variant

In November 2021, a variant of the SARS-CoV-2 coronavirus emerged. It was named omicron by the WHO and listed as a variant of concern. Cases of the highly contagious variant, including a sub-variant called BA.2 (*stealth omicron),* have caused surges of COVID-19, particularly in areas where safety precautions have been relaxed. The omicron variant was responsible for the largest surge since 2019.

Omicron presented a large number of mutations that appeared in one fell swoop. It has over 50 mutations, many in the spike protein, which is how it gets into the human cell in the first place.

However, remember that the spike protein is one of the most prominent exterior features of the virus that our immune system recognizes, responds to, and uses to develop antibodies.

Unfortunately, mutations gave omicron the ability to escape weak immune responses, making it more transmissible from person to person. Omicron and BA.2 do not cause more severe illness. There is evidence that omicron may cause less severe disease than the delta variant. However, omicron may have more affinity for our upper airways than our lower airways, rendering it more easily spread through talking, coughing, or breathing, especially from maskless people. Omicron also may have more potential for airborne aerosolization of the virus, hanging in the air in indoor settings. Omicron, especially its sub-variant, BA.2, is more contagious than the original coronavirus or the delta variant.

Data revealed that being up to date on vaccines and boosters can spare most people from the worst of COVID-19. Being fully vaccinated and getting a booster as soon as you are eligible provides excellent protection against hospitalization and death. In addition, current data suggest that even if you were exposed to omicron or BA.2, being up to date with COVID-19 vaccination and boosters will protect you and prevent severe disease.

Certainly, CDC research findings from January 2022 showed that vaccine boosters effectively prevent severe COVID-19. Therefore, the CDC encourages boosters for eligible people and recommends staying current on all COVID-19 vaccines. For example, during both Delta- and Omicron-predominant periods, receiving a third vaccine dose was highly effective at preventing COVID-19-associated emergency department and urgent care encounters, at 94% and 82%, respectively, while also preventing COVID-19-associated hospitalizations, at 94% and 90%, respectively.

A March 2022 CDC study found that getting two or three doses of an mRNA COVID-19 vaccine was associated with a 90% reduction in risk for COVID-19-associated invasive mechanical ventilation or death. In addition, the protection of three mRNA vaccine doses during omicron predominance was 94%. However, widespread infection with the delta and omicron variants resulted in increased cases in children, including uncommon severe infections and deaths.

The Possibility of More Variants

As long as the coronavirus spreads through the population, mutations will continue to emerge, and the delta and omicron variant families will continue to evolve. The truth is that new variants of the SARS-CoV-2 virus are detected weekly. Most of them will come and go. Some of them will persist but not necessarily become prevalent. Some of them increase in the population for a while, and then they fizzle out. When a change in the population's infection pattern initially appears, it can be extremely difficult to determine what is driving it. It could result from changes to the virus or changes in human behavior. It is, however, worrisome that similar changes to the spike protein are arising independently on multiple continents.

Most of the genetic changes in the coronavirus are like the scars people accumulate over a lifetime, which possess no great significance or functional role. However, when the evidence is strong enough that a viral genetic change is causing a change in the virus's behavior, new insight is gained about the coronavirus. In addition, the virus appears to have some limitations in its evolution, which means there is some hope that variants will not fully escape vaccines.

In any case, updated versions of the current vaccines are being evaluated, while there is no clinical trial evidence yet that variant-specific vaccines would provide significantly greater protection. Though SARS-CoV-2 is changing gradually, it's still much less genetically diverse than influenza. As with any virus, changes must be carefully monitored to ensure that testing, treatment, and vaccines are still effective.

Variants and Safety Precautions

Efforts must be sustained to prevent viral transmission by practicing safety precautions, vaccinating as many people as possible, and encouraging boosters among eligible people. While it is indisputable that vaccines and boosters are the medical miracles of 2020, there is a need to continue re-emphasizing basic public health measures.

The available tools enable individuals to manage risk, including wearing high-quality masks or respirators. Those rated FFP2 or FFP3 are more protective than cloth masks and often easier to wear. Hand-washing and avoiding large indoor gatherings, especially with unmasked people, are other ways to mitigate the risk of infection. Unfortunately, it is within the realm of possibility that other variants could emerge. Immunity seems to decline after several months, but boosters dramatically improve immunity. New variants remind us that we should use multiple tools for our safety and those we care for.

An In-Depth Look at Variants

As mentioned above, viruses constantly change through mutation. A virus with one or more new mutations is called a variant of the original virus. The Centers for Disease Control and Prevention (CDC) identified two variants of the virus (SARS-CoV-2) that causes coronavirus disease 2019 (COVID-19) as variants of concern.

- **Delta (B.1.617.2):** This variant is nearly twice as contagious as earlier variants and might cause more severe illness. The greatest risk of transmission is among unvaccinated people. People who are fully vaccinated can get vaccine breakthrough infections and spread the virus to others. However, vaccinated people spread COVID-19 for a shorter period than unvaccinated people. While research suggests that COVID-19 vaccines are slightly less effective against the delta variant, the Pfizer-BioNTech, Moderna, and Janssen/Johnson & Johnson COVID-19 vaccines still appear to protect against severe COVID-19.

- **Omicron (B.1.1.529) and BA lineages:** The variant B.1.1.529 spreads more easily than the original virus that causes COVID-19 and the delta variant. However, omicron appears to cause less severe disease. Fully vaccinated people can get breakthrough infections and spread the virus to others. But, the COVID-19 vaccines are effective at preventing severe illness. This variant also reduces the effectiveness of some monoclonal antibody treatments. Omicron has a few major offshoots, also termed sublineages, including BA.1, BA.2, BA.3, BA.1.1, and BA.2.12.1. BA.2 made up about 62% of COVID-19 infections that had genetic sequencing in the U.S. during a week in late April of 2021, according to the CDC.

To strengthen protection against COVID-19 and circulating variants, the CDC recommends additional primary shots and booster doses of COVID-19 vaccines in specific instances:

- **Additional primary vaccination:** The CDC recommends an additional shot of an mRNA COVID-19 vaccine for some people with weakened immune systems, such as those who have had an organ transplant. People with weakened immune systems might not develop enough immunity after vaccination with two doses of an mRNA COVID-19 vaccine or one dose of the Janssen/Johnson & Johnson COVID-19 vaccine. An additional shot using an mRNA COVID-19 vaccine might improve their protection against COVID-19. The additional primary shot should be given at least four weeks after the second dose of an mRNA COVID-19 vaccine or one dose of the Janssen/Johnson & Johnson COVID-19 vaccine. The additional primary shot should be the same brand as the other two mRNA COVID-19 vaccine doses that were given. If the brand given isn't known, either brand of mRNA COVID-19 vaccine can be given as a third dose.

- **Booster dose.** A single booster dose is recommended for people aged 12 and older after getting two doses of an mRNA COVID-19 vaccine or one dose of a Janssen/Johnson & Johnson COVID-19 vaccine. If an individual of age 12 or older has been given both doses of the Pfizer-BioNTech COVID-19 vaccine and it's been at least five months, he or she should get a single booster dose. Teens of ages 12 to 17 should only get the Pfizer-BioNTech COVID-19 vaccine booster.

  For people aged 18 or older, the Pfizer-BioNTech or Moderna COVID-19 vaccine boosters are preferred in most situations. If you are 18 years or older, have been given both doses of the Moderna COVID-19 vaccine, and it's been at least five months, you should get a single booster dose. The Pfizer-BioNTech or Moderna COVID-19 vaccine boosters are preferred in most situations. If you are 18 years or older, have been given one dose of the Janssen/Johnson & Johnson COVID-19 vaccine, and it's been at least two months, you should get a single booster dose. The Pfizer-BioNTech or Moderna COVID-19 vaccine boosters are preferred in most situations.

Pregnant women can also get a booster dose. People with a moderately or severely weakened immune system should get an additional primary shot and a booster dose. For example, if you have a weakened immune system, had two doses of an mRNA COVID-19 vaccine and an additional primary shot, and it's been at least three months since the additional shot, get a single booster dose. An mRNA COVID-19 vaccine is preferred. If you have a weakened immune system, had one dose of the Janssen/Johnson & Johnson COVID-19 vaccine and an additional primary shot of an mRNA COVID-19 vaccine, and it's been at least two months since the additional shot, get a single booster dose. An mRNA COVID-19 vaccine is preferred.

- **Second booster dose.** A second booster dose of an mRNA COVID-19 vaccine is recommended for certain people who have a weakened immune system and people aged 50 or older. This second booster dose can be given to those eligible four months after a first booster dose of any authorized or approved COVID-19 vaccine.

Let us look at the science of variants more deeply. A variant means a mutation that occurs in a virus over time. Scientists always knew that mutations or variations in the genome of the coronavirus would occur over time. The RNA genome is more prone to undergoing mutations. As with natural evolution, the virus has the opportunity to replicate its genome as it infects people. Every time it replicates its genome, there are chances for errors or mutations to occur. Since the coronavirus virus has been with us, it has replicated its genome billions of times, and the errors have been incorporated. These have resulted in mutations or variants.

Once again, let us revisit our main concerns with COVID-19 variants. There are three major concerns with any variant:

- Does it spread more easily? In other words, will we start seeing more cases because the virus has gained the ability to be transmitted from one person to another more efficiently?
- Does the mutation in the virus cause it to be more severe in terms of disease or cause higher death rates?
- What are its effects on vaccination and potential therapeutics? As the virus changes over time, does the efficacy of vaccines decline? Are therapies available today less effective against treating the infection when these variants become more prominent? Will current COVID-19 vaccines protect people from the variant strains?

It is safe to say that, for currently circulating variants of SARS-CoV-2, the vaccines will offer some level of protection against those strains. Even when we talk about reduced efficacy, we are not implying going from 95% efficacy to zero. It might reduce efficacy from 95% to 85% against preventing severe COVID-19 disease.

Another vital question is, do COVID-19 tests detect new variants? Most tests look for at least two, sometimes three, different parts of the viral genome. So, if there are mutations in one gene, there's still a backup in the test to detect that variant. For example, testing for SARS-CoV-2, most commonly accomplished by PCR (polymerase chain reaction) testing, will continue to detect these variant strains.

Yet, work is ongoing to continue to sequence these viruses, and updates will be made to the tests to ensure they are still effective in the long term. What can be done to prevent the emergence of new variant strains? Viruses cannot mutate if they don't replicate. There, we need to prevent viruses like SARS-CoV-2 from having the opportunity to replicate. In very simple terms, we need to prevent them from infecting people. The best way we can accomplish that goal is to continue to take current preventive measures. Masking and social distancing, especially indoors and in confined spaces, will always remain effective.

Finally, getting vaccinated is probably the most important measure. Even though a vaccinated person could still be infected with the virus, such a person's viral load will not be as much as that of an unvaccinated person. Therefore, it will be much less likely possible to transmit that virus to other people. Vaccination, masking, and social distancing will continue to be important in helping to drive down the number of infections, which will slow the emergence of new variants.

CHAPTER TEN

A WORLD IN THE THROES OF UNCERTAINTY

The COVID-19 dilemma is far from over. The coronavirus is still mutating. Even at one of the lowest death rates of the pandemic, it is still claiming the lives of hundreds of people daily, killing more than twice the number of deaths in car accidents. Worldwide, its costs are still disproportionately borne by millions of immunocompromised people and healthcare workers who continue to face unsafe working conditions on the frontlines. Moreover, variants of the coronavirus are at the forefront of issues that continue to drive an unrelenting pandemic.

The Delta Variant

The delta variant is concerning because it is spread much more easily than prior versions of COVID-19 that we have seen, meaning that people are still getting sick, hospitalized, and dying from the infection. When the delta variant first reared its head, it was particularly concerning because it seemed more highly transmissible. Yet, the good news has been that the COVID-19 vaccine is still highly protective against getting infected, ending up in a hospital, or dying from the infection.

A breakthrough infection is a rare event where someone fully vaccinated still gets infected with COVID-19. The good news is that these breakthrough infections are generally asymptomatic or only lead to mild illness. As a result, the vaccine remains highly protective against winding up in a hospital or dying from the infection.

The mutations, which caused the delta variant, seem more likely to break through the vaccine immunity than the original types of COVID-19. So, if you are one of those very rare breakthrough cases that we've been hearing about who still gets infected despite being vaccinated, the assurance is that they are rare events. They are not what is driving a surge in cases. The surge is really among unvaccinated people, predominantly young and middle-aged. If a vaccinated person gets infected with COVID-19, he can transmit it to others. Thankfully, vaccines significantly reduce your risk of getting infected, which reduces your risk of passing it on. However, yes, it is possible; there are still studies on how that risk compares with unvaccinated people.

Knowing that in rare cases, a vaccinated person can transmit COVID-19 to someone else is one of the primary reasons behind the recommendation that vaccinated people continue to wear a mask if they are indoors in areas where we see a lot of virus circulation. Guidance on masking has changed with the delta variant for several reasons. One is that the variant is highly transmissible. It is spread much more easily than prior types of COVID-19 that we have seen, so a layered prevention strategy becomes even more important. That is why vaccination and wearing masks will prove to be more protective than using either of them alone.

With the delta variant, we see increased cases among children. The American Academy of Pediatrics reported a significant increase in COVID-19 cases among people under 18 years. That is why children who are eligible to be vaccinated, 12 years or older, should get vaccinated. Vaccination remains the single most effective measure that we will have. Conversely, the best way to protect children who can't be vaccinated, for one reason or the other, is to ensure that everyone around them is vaccinated. That means anyone over 12 years of age should go out and get vaccinated, especially before returning to school after a holiday. Other strategies like wearing a mask when you're out in public and regular handwashing remain important for preventing COVID-19.

Currently, three vaccines are approved for use in the United States. The Pfizer vaccine is approved for children of over 12 years of age. In the early days of the vaccination drive, while the older part of the population was vaccinated, children and young adults made up a larger proportion of new cases of COVID-19. There were various reasons for this: No vaccine has been approved for children under 12 years of age, so they were still very vulnerable to infection. Also, vaccine uptake among young adults was not as high as hoped, so a large proportion of that population remained susceptible to infection.

The Omicron Variant

A lot remains unknown about Omicron, but a worrying trend has become clear. This variant continues to spread rapidly. In South Africa, the UK and parts of continental Europe, and countries that have excellent variant surveillance and high immunity against COVID-19, Omicron cases grew exponentially. The variant outcompeted the already highly transmissible Delta in South Africa and was expected to do the same elsewhere.

According to estimates, every person with Omicron infects 3-3.5 others, which is roughly how fast the coronavirus spread when it first went global in early 2020. In other words, Omicron spreads in highly immune populations as quickly as the original virus did in populations with no immunity. The fear is that big Omicron waves will still lie ahead if this trend continues and is not controlled. Cases were already surging ahead of winter.

At a time the US is still facing a relatively low vaccination rate, Omicron threatens to eat away at the immunity America thought it had. By no means is one saying the pandemic clock has reset back to early 2020. On the contrary, vaccines and previous infections have blunted the worst effects of the coronavirus. Even if protection against infection is eroded, which experts somehow expect, given Omicron's heavily mutated spike protein, protection against severe disease and death should be more durable.

Hospitalizations, rather than cases, might be a better measure of the virus's impact. But if, at any time, cases balloon dramatically, even a tiny percentage of patients becoming seriously ill can turn into too many hospitalizations all at once. Therein lies the danger possible with Omicron. This is because a small proportion of severe diseases will present a very gloomy picture if it is multiplied by millions of cases. But, from the mathematical perspective, a tiny percent of a huge number is still a big number. So, a largely mild but uncontrolled Omicron wave could cause a lot of pain, hospitalizations, and death across a nation and the border of nations.

The ultimate impact of Omicron will depend on how tiny that tiny percent is and how huge that huge number is. Given the early trends out of South Africa, the UK, and Denmark, a large Omicron wave is possible, though not guaranteed. An optimistic note is that the absolute numbers of Omicron cases detected are still small and may be skewed by chance. That means we could be overestimating the variant's growth by specifically searching for it. Yet, Omicron consistently increased in the three countries looking hardest for it. That also means it is likely increasing quietly everywhere else. At the same time, Omicron doesn't appear virulent so far; but this observation comes with even bigger caveats.

Doctors in South Africa, where Omicron was initially predominant, had not seen as many severe cases in previous waves. Other countries with small numbers of Omicron haven't found many sick patients either, but there are several reasons to believe that the news on severity could turn out less rosy than it currently appears. First, we are probably still in Omicron's nascent phase. Infections take weeks to progress to severe infections and eventually to death.

In 2020, the first COVID case in the US was confirmed on January 20, 2020. The first official COVID death was not reported until February 29, 2020. That means the current picture may yet be deceptive. People contracting the virus early in a wave may be disproportionately young and healthy. That is because they are probably taking fewer precautions than elderly persons or those who are immunocompromised. South Africa's population is fairly young, with a median age of 28, compared with 38.5 for the US. Although vaccination rates are low in South Africa, where less than a quarter are fully vaccinated, immunity from the previous infection is very high, possibly up to 62%. A good number of Omicron cases are likely to be reinfections. Cases in young people, those previously infected, or both should be largely mild. If Omicron cases in this population were most severe, that would be catastrophic; the fact that they're not right now is merely a mild cause for relief.

Scientists are working desperately to understand Omicron's effect on vaccinated people. Even if most breakthrough cases continue to be mild in the vaccinated, a small increase in severe cases can still impact hospitalizations by the *tiny percent of a huge number* rule. Ultimately, it is fast appearing as if protection against infection after two doses is not optimal. In real terms, Omicron was a huge leap in the evolution of the coronavirus. However, the picture is not all dismal because protection against severe disease will likely hold up much better against Omicron.

Protection against severe disease tends to be more durable because of how the immune system works. The first-line defenses of neutralizing antibodies might wane, but other slower parts of the immune system, such as T cells, can hold against severe disease. A booster can also strengthen the immune response. At least, current data show that two doses plus infection, or three doses, get people to a higher baseline of neutralizing antibodies, which can better withstand the erosion from Omicron.

Current data show that fewer than half of Americans over 65 have gotten a booster vaccine dose, even though they were prioritized because the elderly people tend to mount weaker vaccine responses. 13 percent are still not fully vaccinated, and 13 percent of the 54 million Americans over the age of 65 translates to 7 million people at risk for requiring hospitalization if they get COVID. The unvaccinated population remains vulnerable to Omicron, as do immunocompromised people who don't mount a good response to the vaccine.

Omicron also arrived on the cusp of the 2021 holiday season, when Americans were gearing up for holiday parties and travels. It was an unfavorable time for a new variant. People had already made their plans, and it was going to be hard to disrupt them. Meanwhile, other seasonal respiratory illnesses, such as the flu, which could also burden hospitals, were rising. Nearly two years of the pandemic had left many hospitals understaffed. Added to that, many healthcare workers were resigning in droves.

A *tiny percentage of a huge number* problem has been present since the beginning of the pandemic. The problem is that while the coronavirus is much less deadly than the epidemic viruses of the past, SARS, MERS, or Ebola, it is still much more transmissible. Across the population, adding up to so many severe cases, it finally overwhelmed the healthcare system. COVID patients got worse care, as did anyone else unfortunate enough to get sick or injured during the big surges.

Yet, if the truth is told, the world is not in the same position as in early 2020 because of the available tools to control Omicron. Also, thankfully for the scientists in South Africa who recognized the risk of the Omicron variant early enough, there is time to put them in place. Vaccines will likely protect against severe infections, and a third shot will likely boost that protection. Manufacturers are working on an Omicron-specific booster. We better understand the virus's airborne transmission and how to stop it with masks and ventilation. We have antivirals on the horizon. We have rapid tests, and they should be easier to get. We know social distancing has curbed the virus before. Omicron is spreading fast, but we know how to slow it down.

The Vagaries of Vaccination

One knows precisely what an individual means if he says he is fully vaccinated. It meant, at the very least, two weeks before, he had received two doses of the Moderna COVID-19 vaccine, two doses of Pfizer, or one dose of Johnson & Johnson. In reality, however, what it means to be vaccinated encompasses much more variety. Some people who have gotten their initial doses haven't gotten a booster dose, and some have mixed and matched the brands of their first shots and their booster. Moreover, everyone is on his timeline, depending on when he gets his shots. Added to this complexity, children under 5 are still ineligible for vaccination. The complexity of vaccination has introduced questions people didn't previously have to deal with: Is it still safe to be exposed to someone vaccinated but has not had a booster dose? Can unvaccinated children safely spend time with vaccinated adults who have not had a booster dose?

When determining whether a gathering will be risky or not, the most important variable is whether attendees are vaccinated. An informed COVID-19 researcher wouldn't feel safe spending time with an unvaccinated person indoors. If a meeting were inevitable, it would be outside and with masks. The next most important distinction is whether someone due for a booster dose has received one. The CDC advises that adults should get booster doses at least six months after their initial doses of Pfizer or Moderna or at least two months after their dose of Johnson & Johnson. A booster dose's full effects take about two weeks to be established.

People who have gotten a booster dose are safer to spend time with, but not having one isn't entirely a disaster. Due to booster doses, the term *fully vaccinated* doesn't indicate much about someone's vaccination history as it used to. However, the CDC's definition of *fully vaccinated* has not changed. It still applies to anyone who received their initial doses, but it doesn't capture whether an adult has received a now-recommended booster dose. Eventually, the Omicron variant or waning vaccine effectiveness could make three doses the new standard for full vaccination. This would add another element of complexity since some *fully vaccinated* people might abruptly revert to the category of *partially vaccinated.*

As people get their booster doses, many get a different brand of vaccine than the one they initially received. Moreover, some brands and combinations appear to be more protective than others. Still, any hierarchy of vaccines is probably not worth obsessing over because the differences between them are the differences between being filthy rich and rich. Indeed, those who can even ponder such distinctions are exceedingly lucky. Much of the world would be thrilled to have access to any vaccines.

In the US, a current irritation is that children younger than 5 are still not eligible for any vaccines, which complicates the risk index for their families. As complicated as vaccination statuses are now, they might become more complicated in the future. The threat Omicron poses is still unclear, but if it turns out to be more contagious, deadlier, and able to evade existing vaccines, an additional dose formulated especially for Omicron might become necessary. At that point, people's vaccine sequences would be even more varied. Some would get their fourth shot, while others would get an Omicron-specific dose as their first.

Yet, the distinctions around booster doses, timing, and brands should fade in importance over time. As the pandemic continues, many people will develop immunity by getting vaccinated or recovering from an infection. For now, the complexity is stressful, but it is a side effect of something good—that more people can get more protection from the virus. The complexity will eventually pass, but thankfully, the advantages of vaccines will not pass.

A Predictable, Yet Perturbing Pattern

Most people are putting the pandemic behind them. Now that cases are crashing, even many who have stayed cautious for two years are talking about returning *to normal* and *living with COVID-19*. This mentality has also translated into government policy. The last pandemic restrictions are fading nationwide. In his most recent State of The Union address, President Joe Biden of the United States declared, "Most Americans can remove their masks, return to work, stay in the classroom, and move forward safely." Other rich, highly vaccinated countries are following the same path. In the UK, for instance, those with COVID-19 no longer have to self-isolate. Of course, it helps that these countries have more vaccine doses than they know what to do with them and a stockpile of tools to test and treat their residents if and when they get sick.

In the third world, COVID-19 is much harder to ignore. More than a year after the start of the mass-vaccination campaign, nearly three billion people are still waiting for their first vaccine dose. While an average of 80 percent of people in high-income countries have gotten at least one dose, that figure stands at just 13 percent in low-income countries. Moreover, virtually no booster doses have been administered in the poorest countries. Such low vaccination rates are taking their toll.

Although the official death count in India is about 500,000, the reality might be closer to a grim five million excess deaths; most of those deaths happened after vaccines rolled out in the Western world. The rush in the wealthier countries to declare the pandemic *over* while it ravages the underdeveloped and underserved world is predictable. In fact, the same trend has played out repeatedly.

Infectious diseases such as malaria, tuberculosis, and AIDS, now seen as Third World diseases, were once serious threats in wealthier countries. With COVID-19, the developing world has again been left to fend for itself against an extremely transmissible virus attack without the necessary vaccine doses, tests, and treatment tools. Some pandemics never truly end; they just become invisible to people in the developed world.

We know malaria as an infectious disease that affects poor *tropical* countries, but for several thousands of years, malaria was a global menace. During the 20th century, the disease was estimated to have accounted for up to 5% of all human deaths. It was eradicated from the developed world by the 1970s, but the rest of the world was left behind. In 2020, there were an estimated 240 million malaria cases; and nearly all 627,000 deaths occurred in sub-Saharan Africa. Unfortunately, for a disease that affected even our neolithic ancestors, the world had to wait until *2021* for the first-ever malaria vaccine. The World Health Organization recently endorsed this partially-effective malaria vaccine, but expanded manufacturing and scale-up plans remain unclear.

A similar phenomenon unfolded with tuberculosis—a disease so old that the DNA of TB bacteria have been identified in Egyptian mummies. *Consumption,* as TB was once called, was highly prevalent in Europe and North America. From the 1600s to the 1800s, TB caused 25% of all deaths in Europe. By the 1980s, TB case numbers had decreased significantly in the West, largely due to drug treatments and reductions in poverty. But again, TB remains a problem in developing countries. In 2020, TB killed 1.5 million people, more than 80% of whom lived in low- and middle-income countries. Investments and innovations to make the disease less devastating have been scarce. For example, the TB vaccine we use today is more than 100 years old and has limited efficacy in adults.

Unlike malaria and tuberculosis, HIV/AIDS was identified 40 years ago, and we are seeing the same trend. After the disease emerged in the early 1980s, it went from a condition thought to affect only gay men in the developed world to a global pandemic that mostly affects the underdeveloped world today. In 2020, nearly 38 million people globally were living with HIV, and 680,000 people died from AIDS-related illnesses, with two-thirds of both cases and deaths in Africa.

When effective antiretroviral drugs first became available in the early 1990s, they were expensive and mainly accessible to people in high-income countries. For these lifesaving tools to reach the third world took a tremendous amount of near-heroic activism and years of effort, and millions of people, mostly Africans, died due to inaction. Even today, there is as yet no vaccine against AIDS.

Despite the continued toll of these *big three* infectious diseases, they are rarely spoken of as pandemics. A pandemic means a pandemic that no longer kills people in rich countries, while an endemic means a disease the world could get rid of but hasn't. HIV/AIDS, TB, and malaria are pandemics that have been defeated in rich countries. Allowing them to persist elsewhere is a policy choice and a budgetary decision.

With the coronavirus, the underdeveloped world is being left behind once again; wealthier countries are rapidly losing interest. Suppose the virus continues to fizzle out in these areas. In that case, they might show less urgency in sharing vaccines and other resources, stop investing in new products to fight the virus, and place the burden of disease control primarily on resource-strapped, low-income countries.

Wealthy countries such as the United States have donated hundreds of millions of doses to the COVID-19 Vaccines Global Access Facility (COVAX). Still, citing supply issues, the initiative didn't get halfway to its goal of delivering 2 billion doses last year. Even if rich nations continue to offer charity and donations, they seem less likely to support efforts that would let lower-income countries procure and manufacture their tools to battle the coronavirus. HIV medication became affordable to the third world only when countries such as India started manufacturing generic drugs. The same must happen for COVID-19 vaccines to be more accessible.

After Omicron, some have suggested that it is too late to meet the WHO's target of vaccinating 70 percent of the world by mid-2022. When we should be redoubling efforts to increase vaccination, the narrative that it's too late to vaccinate the world could have a chilling effect on the global COVID-19 vaccination campaign. Yet, the developed world is repeating its mistakes again, which will have devastating consequences for billions of people worldwide. Diseases becoming *endemic* ought not to be coded for inaction or lack of consideration for those with fewer resources and many vulnerabilities. Even when they're invisible to some, high death and infection rates cannot be seen as acceptable or normal.

For now, the biggest problem with the developed world proclaiming that the coronavirus pandemic is *over* is that it manifests the *opposite* outcome. Eventually, even wealthy countries will bear the burden of becoming complacent about COVID-19. Allowing infectious diseases to circulate in any part of the world within large populations of unvaccinated people will surely result in the emergence of new variants that will affect all nations. Privileged people should not decide on their own that a global pandemic is over. The way out is the same as it has always been—to ensure everyone arrives at the finish line, not just a few selected. Humanity did this with smallpox and could soon achieve this with polio and guinea-worm infections. The real barrier to ending the COVID-19 crisis around the world is not science or resources; it is elitist complacency.

Hope in the Horizon

US health regulators recently cleared the use of a COVID-19 medicine from Pfizer Inc., making it the first drug that newly infected patients can now take at home to stay out of the hospital. The US Food and Drug Administration authorization permits doctors to prescribe the medicine to high-risk patients of age 12 and older early in the course of the disease, shortly after they develop symptoms.

The FDA said the drug, named Paxlovid, works safely and should be an important tool in reducing hospital admissions and death. The medicine arrives as the Omicron variant sweeps across the

United States. Researchers expect antivirals like Paxlovid to be effective against the new strain because of how they work, but they are waiting for studies to confirm that optimism. Nevertheless, the authorization is a major step forward in the fight against the coronavirus. Paxlovid is expected to be available soon at pharmacies and hospitals.

Naturally, initial supplies will be limited until Pfizer can ramp up production. The US government plans to initially spend $5.29 billion to purchase 10 million treatment courses, and the drug will be free to patients. Pfizer Inc. said it aims to make 120 million treatment courses in 2023, up from its previous projection of 80 million. The medicine's key advantage over most existing treatments is that patients can easily take them at home, unlike antibody therapies or the antiviral Veklury that usually require administration at a hospital.

The medicine is for the newly infected who are considered at high risk of severe disease, which the Centers for Disease Control and Prevention has said include people of 65 years old and above or those with other medical conditions like cancer, diabetes, or chronic lung disease. Paxlovid's launch puts Pfizer in a position to have a commanding presence in the world's response to the pandemic. Pfizer Inc. said laboratory tests indicate that Paxlovid could work against the Omicron variant because the drug works differently from other therapies targeting the spike protein on the virus.

Researchers designed the medicine to stop the coronavirus from spreading in people by blocking the activity of a key enzyme, protease, which the virus needs to replicate. The enzyme isn't believed to be mutated in Omicron, so Pfizer Inc. and other researchers say the antiviral medicine will likely work against Omicron. The authorization came after researchers found that Paxlovid was safe and would cut the risk of hospitalization or death by about 89% if people at high risk of severe COVID-19 took the medicine within three days of diagnosis.

Some researchers have raised safety concerns because Paxlovid is taken with another drug. The second drug, called ritonavir, helps Paxlovid fight the virus longer but can cause side effects when taken with certain widely used heart and cholesterol medicines. Under the authorization, the FDA said Paxlovid is contraindicated in patients using certain drugs for conditions including cancer, irregular heart rhythms, and depression. The FDA also recommended against giving Paxlovid to patients with severe kidney disease or liver impairment and said patients with moderate kidney issues should take a reduced dose. "In patients with moderate renal impairment, a reduced Paxlovid dose is needed," FDA said.

In reviewing Pfizer's application for Paxlovid clearance, European health regulators recently issued preliminary guidelines recommending that people shouldn't use it if they are pregnant or trying to get pregnant or if they are taking certain medicines or have severely reduced kidney or liver function. However, Pfizer has said that it expects doctors and pharmacists to find most of the drug interactions manageable.

Under the authorization, patients aged 12 and older should begin a course of treatment within several days of symptoms emerging to prevent their cases from turning serious and requiring hospitalization. People with a prescription should take two Paxlovid pills and one ritonavir pill twice daily for five days. Pfizer has licensed Paxlovid's formula so that generic drugmakers can make it for low- and middle-income countries. Pfizer is also studying the drug in a separate study in adults at lower risk of developing severe disease, including vaccinated individuals.

EPILOGUE

WHAT THE FUTURE HOLDS

Out of It, Or Not Out of It?

Anthony S. Fauci is the chief medical adviser to the 46th president of the United States, Joseph R. Biden Jr. In optimistic glee, sometime in the first quarter of 2022, he said, *"The United States is finally out of the full-blown explosive pandemic phase that has led to nearly 1 million deaths from COVID-19 and more than two years of suffering and hardship. We're really in a transitional phase, from a deceleration of the numbers into hopefully a more controlled phase and endemicity."*

He clarified that the pandemic is not over and that the United States could still see new waves of infections as the virus mutates and spins off highly transmissible variants. Yet, Fauci and other infectious-disease experts hope that the population has built up enough immunity from previous infections and vaccinations to avoid another devastating surge in hospitalizations and deaths.

"The world is still in a pandemic. There's no doubt about that. Don't anybody get any misinterpretation of that. We are still experiencing a pandemic," Fauci further explained. His comments followed a report from the Centers for Disease Control and Prevention indicating that

roughly three in five people in the United States had already been infected by the coronavirus. In addition, about one in four people had a first-time infection during the winter wave of 2021 caused by the Omicron variant. Those startling numbers suggested that the US has a much higher level of collective immunity than it had before the arrival of the Omicron variant. How long that immunity will persist and to what extent new coronavirus variants can evade it is far less clear.

By the first quarter of 2022, the omicron sub-variant BA.2.12.1 had become the latest coronavirus version to arrest public health experts' attention. It rapidly gained traction, with the CDC estimating that it was already accounting for about 30% of new infections. It is about 25 percent more transmissible than the omicron sub-variant BA.2, which is more transmissible than the original omicron strain. The virus's ability to mutate and work around the human immune system has made experts cautious about predicting the trajectory of the pandemic.

Fauci's optimism was faulted. He had to take further pains to explain his thinking. "The country has not reached a major milestone," he said. He was pointing out what is clear from the coronavirus trackers in recent months: the country is in a different position than it was. He said the United States was initially in the *full-blown pandemic phase* and then entered a period he referred to as the *deceleration* phase.

"There's the full-blown pandemic dynamic, the way we were months ago, where we were having 900,000 cases a day, tens of thousands of hospitalizations, 3,000 deaths a day," he continued, "The deaths went from 3,000 down to 300." He also mentioned that the country is transitioning to the *control* phase. That is where he expects the health emergency to settle, with the virus circulating but not causing devastating new waves of hospitalization and death.

One thing is certain. The United States is now firmly in what might be whimsically termed the ***figure it out yourself*** era of the COVID-19 pandemic. Across the country, almost all government efforts to curtail the coronavirus have evaporated. Mask mandates have been lifted on public transit. COVID funding remains stalled in the US Congress, jeopardizing supplies of tests and vaccines.

The White House and the CDC have framed COVID as a problem for individuals to act upon. Still, action is difficult when cases and hospitalizations are underestimated, many testing sites have been shut down, and previously reliable guidelines downplay the continued spread of the coronavirus. But, the COVID question is far from resolved. The coronavirus is still mutating. Even at one of the lowest death rates of the pandemic, it still claims the lives of hundreds of Americans daily, killing more than twice as many people die in car accidents.

For more than a year, the US has focused on using vaccines and drugs to avert severe disease and death while deprioritizing other means of preventing *infections*, such as masks and ventilation. To a degree, this strategy is working. Cases and hospitalizations recently spiked again, while ICU admissions rose gently and deaths have remained stable. Yet, *infections still matter* and affect all of American society, including the vaccinated. In addition, the coronavirus periodically takes waves of healthcare workers out of action, with the entire healthcare system perpetually overburdened and unable to provide its former standard of care.

People are still being disabled by long COVID-19, often without ever being hospitalized. Uncontrolled infections are a gift to the virus, which keeps birthing new variants that could prolong the current crisis level or send it spiraling back into a greater level of disruption. Many sensible policies, for instance, mask mandates, seem unlikely in the current political climate. It is a dismal scenario.

A Dismal Picture

Sadly, the most infectious and transmissible variant arrived two-and-a-half years from the beginning of the coronavirus pandemic. Serial COVID-19 waves have left millions of people dead, and only vaccines have helped to blunt the toll. As this book was about to go to press, the virus started spreading again. It is evolving, escaping immunity, and driving an upsurge in cases and hospitalizations. The latest version of the coronavirus, BA.5, clearly shows that the pandemic is far from over. According to the World Health Organization (WHO), the newest offshoot of Omicron and a closely related variant, BA.4, are fueling a global surge in cases.

In Europe, the Omicron sub-variants are powering a spike in cases of about 25%, though that number may be higher, given the *almost collapse in testing*. BA.5 is on the march in China, raising anxiety that major cities may soon re-enforce strict lockdown measures that were recently lifted. The same variant has become the dominant strain in the United States, where it recently accounted for 65% of new infections, according to the Centers for Disease Control and Prevention (CDC).

The WHO's Emergency Committee has said that COVID-19 remains a Public Health Emergency of Concern (PHEC)—its highest level of alert, first declared on January 20, 2020, amid rising cases, ongoing viral mutation, and increasing pressure on already overstretched health systems. In a statement, the committee, made up of independent experts, highlighted challenges to the ongoing global COVID-19 response, including a drop-off in testing and spotty genome sequencing, raising the question of just how accurately any nation might reasonably be able to monitor BA.5.

Official data appears to be dramatically undercounting the actual number of infections in the US, leaving the nation with a critical blind spot as the most transmissible coronavirus variant yet takes hold. Some experts think there could be as many as one million new infections daily in the broader US population. Second booster doses appear to be the only solution in sight. Adults who are up to date with vaccinations are less likely to be hospitalized than those who are unvaccinated. Still, only about one in four adults, in the US, who are over 50 years have gotten their recommended second boosters, according to data collected by the CDC.

US health officials are urgently working on a plan to allow second COVID-19 booster doses for all adults amid fears that younger adults' immunity may be waning as COVID-19 cases rise with the dominance of BA.5. Dr. Eric Topol, a cardiologist and professor of molecular medicine at Scripps Research, did not mince words in describing BA.5. "BA.5 is the worst version of the virus we have seen. It takes immune escape, already extensive, to the next level, and, as a function of that, enhanced transmissibility," he said. In other words, BA.5 can easily evade immunity from previous infections and vaccines, increasing the risk of reinfection. Though the variant does not appear to lead to more severe illness, given the extent of BA.5's immune evasion, there will be an escalation in hospitalizations, as seen in Europe and elsewhere where the variant has taken root.

Public health experts in the US may take some solace from the trajectory of the variant in Europe. While many European countries are experiencing a spike in hospitalizations, what is not seen is an increase in intensive care unit admissions. That means the vaccines are still working, but those gaps in immunity are causing the problem. Yet, steep reductions in COVID-19 surveillance worldwide hinder epidemiologists' efforts as they race to trace the virus's evolution.

WHO's Director-General, Tedros Adhanom Ghebrevesus,

said, "Sub-variants of Omicron, like BA.4 and BA.5, continue to drive waves of cases, hospitalization, and death around the world. Surveillance has reduced significantly, including testing and sequencing, making it increasingly difficult to assess the impact of variants on transmission, disease characteristics, and the effectiveness of counter-measures. New waves of the virus demonstrate again that the pandemic is nowhere near over."

Summarily, the musical chair of variants will not end anytime soon unless we do something about it. After two-plus years of erupting into different peaks, the American coronavirus-case curve has the new topography of a long, never-ending plateau. Waves are now so frequent that they collide and uplift like tectonic plates. With high cases still drastically undercounted and hospitalizations going up, this is a disconcerting place to be. The sub-variants keep coming; immunity remains formidable against severe disease but porous to infection and the resulting chaos. Some people get the virus for the first time; others for the second, third, or more, occasionally just weeks apart.

Coronavirus positive test trends are not encouraging, and it is all about the speed of the virus. SARS-CoV-2, as represented by the Omicron clan, is now spewing out globe-sweeping sub-variants at a breathless pace. In the United States, the fall of BA.2 and BA.2.12.1 have overlapped so tightly with the rise of BA.5 that the peaks of their surges have coalesced into one. This means a new, ominous kinsman, BA.2.75, appears to be currently emerging in several parts of the world.

Sadly, public health and policy countermoves are sluggish at best. Pathogens don't spread or transform without first inhabiting hosts. Still, with masks, distancing, travel restrictions, and other protective measures almost eradicated, we have given the virus every opportunity to continue to spread and mutate. More variants mean more infections and vice versa; it is a vicious cycle. It is true that, compared with earlier in the pandemic, hospitalization and death rates remain relatively low. Yet, a high infection rate keeps us in the vicious viral-evolution cycle. Our stark failing is the unchecked transmission. We might be ready to return to normal and forget that the virus exists, but without doing something about infection, we cannot slow the COVID treadmill we find ourselves on.

The speed at which a virus mutates hinges on two main factors: the microbe's inherent capacity for change and the frequency with which it interacts with hospitable hosts. Coronaviruses don't tend to mutate as quickly as other RNA viruses. For around the first year of the pandemic, SARS-CoV-2 remained with that stereotype, doing roughly two mutations a month, but then came along Alpha, Delta, Omicron, and its many sub-variants. Also, SARS-CoV-2 began to outstrip the abilities of flu viruses to birth versions of themselves that vaccinated and previously infected bodies can't easily recognize.

With most of the world now at least partially protected against the virus, thanks to vaccines and infections, immune evasion is the only way a new variant can spread. Moreover, because even well-defended bodies have not been able to prevent infection and transmission fully, SARS-CoV-2 has had ample opportunity to invade and find genetic combinations that help it maneuver its way around safeguards. That same modus operandi sustains flu viruses, which repeatedly reinfect individuals. It has also defined the Omicron oligarchy, and the longer the Omicron dominance continues, the more difficult it will be for another variant to take over its throne. However, it is unclear why this particular variant has managed to sustain a monopoly. Whatever the secret for its staying power, Omicron has spread far and wide.

To slow down SARS-CoV-2, we, the virus's hosts, could give it fewer places to reproduce by boosting immunity and curbing infections. Unfortunately, the world is nowhere near where it should be on the immunity front, and infections will continue. Only *vaccinations* can shore up our defenses. Billions of people have now received at least one dose of a COVID-19 vaccine, but there are still large pockets of individuals, especially in low-income countries, who have had none. Even among the vaccinated, far too few have had the optimal vaccination doses necessary to prevent infection. Precautionary policies are gone; everyone wants some sense of normalcy—many people who consider themselves COVID-conscious have picked up old social habits again; the floodgates of complacency have opened up; the global situation *has* immensely improved.

Vaccines and treatments have slashed the proportion of people ending up seriously sick and dead, even when case rates climb. The virus's onslaught should continue to soften as global immunity grows. SARS-CoV-2 will eventually, like flu and other coronaviruses, adhere to some form of seasonality, becoming a threat that can be managed with an annually updated vaccination. But, improvement in the COVID situation is not guaranteed, and the current burden of infection remains unsustainably heavy. Long COVID still looms, while mild sickness can leave people bedridden for days and take them away from school, family, and work. Moreover, with transmission rates as high as they currently are, the next variant may arrive even sooner and end up more severe. In simple terms, the world is in a difficult situation.

Pandemic Warning

As I approach the closing pages of this book, I am constrained to address certain vital questions. First, were we forewarned about the current pandemic, and if yes, why did we fail to heed such warning? Second, do we have a pandemic warning infrastructure, or do we need to create one? The world has dealt with several pandemics in the past, including SARS, the first major pandemic of the 21st century. Therefore, COVID-19 is neither the first nor will it be the last pandemic.

US President Donald Trump expressed surprise that the coronavirus would badly affect the United States. He was ill-prepared to accept the fact of COVID-19's existence. Worse, he believed the virus would disappear in a few days. He said that it was unexpected. However, the truth was far from that. Pandemic warnings have been issued at regular intervals in the recent past. Leading experts from the field of public health, as well as global influencers like Bill Gates, had given pandemic warnings in the past, pointing out the under-preparedness of the American government to deal with a pandemic.

There is a recurring trend. The US government and international health agencies spring into action whenever a global health crisis occurs. Then, when the crisis resolves, they replace urgency with complacency, as if nothing had happened and nothing would happen in the future. Funding for healthcare decreases, life goes on as usual, and the crisis fades from memory soon.

The world appears to have forgotten the Spanish flu, SARS, MERS, and swine flu. The most recent warning, a bipartisan report by the Center for Strategic and International Studies, was published on November 18, 2019. That report's prime recommendation stated, "Restore health security leadership at the White House National Security Council." However, a careful review of past reports, books, planning documents, Congressional hearings, speeches, and public testimonies renders it clear that all the challenges of the current COVID-19 crisis have been envisioned over the past 15 years.

There has been a collective failure of world leaders, industrialists, health experts, policymakers, and economists. The machinery for pandemic preparedness comes alive only when the pandemic arrives and goes into a state of hibernation, only to be rudely jolted from sleep when the next pandemic strikes. Then, tragically, the pandemic strikes like a tsunami and kills or maims every susceptible person in its path.

The Global Preparedness Monitoring Board (GPMB), an independent monitoring and advocacy body, urges political action geared towards preparing for and mitigating the effects of global health emergencies. The 15-member board comprises political leaders, heads of agencies, and experts, led jointly by Dr. Gro Harlem Brundtland, former prime minister of Norway, director-general of the WHO, and the secretary-general of the International Federation of Red Cross and Red Crescent Societies.

GPMB, in its September 2019 report, issued a dire warning about the pandemic threat saying, "While disease has always been part of the human experience, a combination of global trends, including insecurity and extreme weather, has heightened the risk. Disease thrives in disorder and has taken advantage—outbreaks have been on the rise for the past several decades and the specter of a global health emergency looms large. If it is true to say *what's past is prologue*, then there is a very real threat of a rapidly moving, highly lethal pandemic of a respiratory pathogen killing 50 to 80 million people and wiping out nearly 5% of the world's economy. A global pandemic on that scale would be catastrophic, creating widespread havoc, instability, and insecurity. The world is not prepared."

Factors Aiding the Emergence of Pandemics

Human beings are largely responsible for pandemics. In our interconnected world, everyone is responsible for creating a pandemic-prone environment. These are likely factors:

- The world's population is speeding towards 9.7 billion by 2050, and more people are expanding into new frontiers.
- Unplanned urbanization and the rapid and mass movement of populations are forcing more people into overcrowded and unsanitary living conditions.
- Global warming allows disease vectors like mosquitoes and ticks to thrive over greater geographic areas.
- Insecurity and conflict are increasing throughout the world.
- Human populations continue encroaching on animal territories, and animal pathogens come in contact with human beings.

Preventing and Mitigating Pandemics

The responsibility for preventing pandemics is a collective one. Here are strategies for preventing and mitigating pandemics:

- The world needs to establish systems for early detection and control.
- Reduction in global poverty should be a priority.
- Sustainable development goals must be achieved by carefully using non-renewable resources and maintaining an ecological balance.
- Environmental degradation has to be avoided at all costs.
- Leaders at all levels should be involved in devising strategies for pandemic prevention.
- The governments of all countries must commit to and invest money into pandemic preparedness.
- Governments should build strong systems involving all agencies and stakeholders to prepare an effective mechanism for preparedness to deal with outbreaks.

- The IMF and World Bank need to help developing countries in their efforts to develop their healthcare infrastructures.
- Donors and philanthropists should unite to establish funding platforms to help the poor and the vulnerable.

Public health is essentially the science of protecting and improving the health of people and their communities. It is concerned with protecting the health of entire populations. Such populations can be as small as a county, as big as an entire country, or even a world region.

The next critical challenge is establishing and maintaining an effective global public health system. The COVID-19 pandemic has refocused the attention of governments and policymakers on the compelling need for seamless universal healthcare through a global public healthcare system.

In a media briefing on May 6, 2020, the WHO said that the coronavirus crisis had revealed the importance of national and subnational health systems. WHO's Director-General, Dr. Tedros Adhanom Ghebreyesus, said those systems comprise global health security's foundation. He further said, "Strong and resilient health systems are the best defense, not only against outbreaks and pandemics but also against the multiple health threats that people around the world face every day." National public health services play a major role in responding to the COVID-19 pandemic. They provide essential services to the government, policymakers, healthcare system, and general public. They also spring into action the moment an outbreak occurs.

With the help of experienced epidemiologists and a trained field force, public health experts operate at the forefront of tackling a public health crisis. They accomplish this objective through door-to-door surveys, contact tracing, laboratory testing, epidemiological surveillance, and analysis while gathering information, which they collate and provide to the government and policymakers. The process also allows them to identify vulnerable groups and provide vital, life-saving information to the general public. Depending on the outbreak's severity, they also assist the government in developing surge capacity in healthcare infrastructure.

Unfortunately, however, in many parts of the world, this previously efficient and effective public health infrastructure has gone into a state of decay due to decades of government neglect. As a result, the world has moved towards managed care, as exemplified by the US insurance-based healthcare system, rather than government-backed public healthcare. The COVID-19 pandemic made everyone sick simultaneously, and in this collective fight, almost everybody has realized that private and managed healthcare is not enough. Had it been enough, morbidity and mortality would not have been so high in the US and European countries.

Ironically, the US, the pioneer of private or managed healthcare, has suffered the most in the COVID-19 pandemic. The death rate of Black Americans and Hispanic Americans is proportionately much higher than that of white Americans. They were the most affected because they are the poorest, often work in essential services, and do not enjoy the benefit of insurance-based healthcare. The COVID-19 pandemic has highlighted the glaring shortcomings and limitations of insurance-based private healthcare. Having said all that, the one indisputable fact is that the world needs to wake up from its slumber. To be forewarned is to be forearmed. A word, as the saying goes, is enough for the wise.

John Uyanne, MD
Los Angeles, California
July 2022

REFERENCES

OVERVIEW

1. WHO. "Naming the coronavirus disease (COVID-19) and the virus that causes it." 2020.

   www.who.int/emergencies/diseases/novel-coronavirus-2019/technical-guidance/naming-the-coronavirus-disease-(covid-2019)-and-the-virus-that-causes-it

2. IRA M. LONGINI, JR., PAUL E. M. FINE, STEPHEN B. THACKER. "Predicting the global spread of new infectious agents." *American Journal of* Epidemiology 123(1998):383-391.
3. David M. Morens, M.D., Peter Daszak, Ph.D., and Jeffery K. Taubenberger, M.D., Ph.D. "Escaping Pandora's Box – Another novel coronavirus." *New England Journal of Medicine* Last modified April 2, 2020.

   www.nejm.org/doi/full/10.1056/NEJMp2002106

4. WHO. "Coronavirus disease (COVID-19) situation report — 116." Last modified May 15, 2020

   www.who.int/docs/default-source/coronaviruse/

situation-reports/20200515-covid-19-sitrep-116.pdf?sfvrsn=8dd60956_2

5. WHO. "Rolling updates on Coronavirus disease." Last modified July 31, 2020

   www.who.int/emergencies/diseases/novel-coronavirus-2019/events-as-they-happen

6. Shayan Sardarizadeh and Olga Robinson. "Coronavirus: US and China trade conspiracy theories." *BBC*. Published April 26, 2020.

   www.bbc.com/news/world-52224331

7. "COVID-19: Global backlash builds against China over coronavirus." *The Times of India*

   https://timesofindia.indiatimes.com/world/china/covid-19-global-backlash-builds-against-china-over-coronavirus/articleshow/75531645.cms

8. Cooper S. "Why is the World Health Organization accused of mishandling the coronavirus pandemic?" *Global News*. Last modified April 16, 2020.

   https://globalnews.ca/news/6826415/world-health-organization-accused-of-mishandling-coronavirus-pandemic-covid-19/

9. Jeff Mason and Paulina Duran. "Donald Trump stops US funding to WHO, blames it for COVID-19 spread." *The Wire*. Published April 15, 2020

   https://thewire.in/world/donald-trump-us-who-funding-covid-19

10. Maron DF. "'Wet markets' likely launched the coronavirus. Here's what you need to know." *National Geographic*. Published

April 15, 2020.

www.nationalgeographic.com/animals/2020/04/coronavirus-linked-to-chinese-wet-markets/

CHAPTER ONE

1. "Report of the WHO-China joint mission on coronavirus disease 2019 (COVID-19)." *WHO.*

   www.who.int/docs/default-source/coronaviruse/who-china-joint-mission-on-covid-19-final-report.pdf

2. Livingston E, Bucher K. "Coronavirus disease 2019 (COVID-19) in Italy," *JAMA* 2020;323(14):1335 published online March 17, 2020, doi:10.1001/jama.2020.4344

3. Fraser C, Riley S, Anderson RM, et al., "Factors that make an infectious disease controllable," *Proceedings of National Academic Science.* 101(2004): 6146–6151

   https://doi.org/10.1073/pnas.0307506101

4. Scott ME, Smith G, eds. "*Parasitic and Infectious Diseases: Epidemiology and Ecology*," San Diego: Academic Press, 1994.

5. Yaeger RG, Protozoa: Structure, Classification, Growth, and Development, Medical Microbiology, 4th edition.

   www.ncbi.nlm.nih.gov/books/NBK8325/

6. Cascella et al., *Features, Evaluation and Treatment Coronavirus (COVID-19).* National Library of Medicine, Last updated June 30, 2022.

7. Chan JF, Kok KH, Zhu Z, et al. "Genomic characterization of the 2019 novel human-pathogenic coronavirus isolated from a patient with atypical pneumonia after visiting Wuhan, Emerging Microbes & Infections, 2020, 9(1): 221–236. Doi: 10.1080/22221751.2020.1719902

8. Wan Y, Shang J, Graham R, et al. "Receptor recognition by

novel coronavirus from Wuhan: An analysis based on decade-long structural studies of SARS." *J Virology* (2020) published online January 29, doi:10.1128/JVI.00127-20

9. Huang C, Wang Y, Li X, et al., Clinical features of patients infected with 2019 novel coronavirus in Wuhan, China, Lancet, 395(10223): 497–506, January 24, 2020; Chen N, Zhou M, Dong X, et al., Epidemiological and clinical characteristics of 99 cases of 2019 novel coronavirus pneumonia in Wuhan, China: A descriptive study, Lancet, 395(10223): 507–513, January 30, 2020; Wang D, Hu B, Hu C, et al., Clinical characteristics of 138 hospitalized patients with 2019 novel coronavirus-infected pneumonia in Wuhan, China, JAMA, 2020; 323(11): 1161–69, February 7, 2020
10. Wu Z, McGoogan JM, Characteristics of and important lessons from the coronavirus disease 2019 (COVID-19) outbreak in China: Summary of a report of 72 314 cases from the Chinese Centers for Disease Control and Prevention, JAMA, 2020; 323(13): 1239–1242, February 24, 2020
11. Severe outcomes among patients with coronavirus disease 2019 (COVID-19) – United States, February 12–March 16, 2020, CDC, Morb Mortal Wkly Report (MMWR), March 18, 2020.

    www.cdc.gov/mmwr/volumes/69/wr/mm6912e2.htm
12. Report of the WHO-China joint mission on coronavirus disease 2019 (COVID-19)

CHAPTER TWO

1. Tan SY, Tatsumura Y. 2015. "Alexander Fleming (1881–1955): Discoverer of penicillin." *Singapore Medical Journal* 56(7): 366–367
2. Porter JD, McAdam KP. 1994. "The Re-emergence of

Tuberculosis." *Annual Review of Public Health* 15: 303–323

3. Johnson NPAS, Mueller J. 2002. "Updating the accounts: Global mortality of the 1918–1920 'Spanish' influenza pandemic." *Bulletin of the History of Medicine*76(1): 105–115
4. WHO. "Report on global surveillance of epidemic-prone infectious diseases – Influenza."

   www.who.int/csr/resources/publications/influenza/CSR_ISR_2000_1/en/
5. Fujimura SF. "Purple death: The great flu of 1918."

   www.paho.org/English/DD/PIN/Number18 article5.htm
6. Tobi S. "Inside the swift, deadly history of the Spanish flu pandemic." National Geographic.

   www.nationalgeographic.com/history/magazine/2018/03-04/history-spanish-flu-pandemic/
7. Tumpey T.M., Baster C.F., Aguilar P.V., et al. 2005 "Characterization of the reconstructed 1918 Spanish influenza pandemic virus." *Science* 310(5745): 77–80
8. Wang MD, Jolly AM. 2004. "Changing virulence of the SARS virus: The epidemiological evidence." *Bulletin of the World Health Organization* 82(7): 547–548
9. CDC. "Ebola outbreak in West Africa."

   www.cdc.gov/vhf/ebola/history/2014-2016-outbreak/index.html
10. CDC. "Ebola Virus Disease (EVD) information for clinicians in U.S. healthcare settings."

    www.cdc.gov/vhf/ebola/clinicians/evd/clinicians.html
11. CDC. "Prevention and vaccine."

www.cdc.gov/vhf/ebola/prevention/index.html

12. Méndez N, Oviedo-Pastrana M, Mattar S, et al. 2017. "Zika virus disease, microcephaly and Guillain-Barré syndrome in Colombia: Epidemiological situation during 21 months of the Zika virus outbreak, 2015–2017." *Arch Public Health*, 75: 65

CHAPTER THREE

1. WHO. "Coronavirus disease (COVID-19) situation report – 116."

   www.who.int/docs/default-source/coronaviruse/situation-reports/20200515-covid-19-sitrep-116.pdf?sfvrsn=8dd60956_2

2. The Guardian. "The worst of COVID-19 is yet to come warns World Health Organization – video."

   www.theguardian.com/global/video/2020/apr/21/the-worst-of-covid-19-is-yet-to-come-warns-world-health-organization-video

3. WHO. "Critical preparedness, readiness and response actions for COVID-19."

   www.who.int/emergencies/diseases/novel-coronavirus-2019/technical-guidance/critical-preparedness-readiness-and-response-actions-for-covid-19

4. MIRROR. "Wuhan shrimp seller may be coronavirus 'patient zero' and says bug 'came from toilet'."

   www.mirror.co.uk/news/world-news/wuhan-shrimp-seller-coronavirus-patient-21767997

5. BBC. "Who is 'patient zero' in the coronavirus outbreak?"

   www.bbc.com/future/article/20200221-coronavirus-the-harmful-hunt-for-covid-19s-patient-zero

6. BBC. "Li Wenliang: Coronavirus kills Chinese whistleblower doctor." Published February 7, 2020

www.bbc.com/news/world-asia-china-51403795

7. BBC. "Li Wenliang: Coronavirus death of Wuhan doctor sparks anger". Published February 7, 2020

www.bbc.com/news/world-asia-china-51409801

8. Josephine Ma. 2020. "Coronavirus : China's first confirmed COVID-19 case traced back to November 17." *South China Morning Post.*

9. Xinhua. "Xinhua headlines: Chinese doctor recalls first encounter with mysterious virus." Published April 16, 2020.

www.xinhuanet.com/english/2020-04/16/c_138982435.htm

10. WHO. "Coronavirus disease 2019 (COVID-19) situation report – 71."

www.who.int/docs/default-source/coronaviruse/situation-reports/20200331-sitrep-71-covid-19.pdf?sfvrsn=4360e92b_8

11. WHO. "Coronavirus disease (COVID-19) situation report – 116."

www.who.int/docs/default-source/coronaviruse/situation-reports/20200515-covid-19-sitrep-116.pdf?sfvrsn=8dd60956_2

12. WHO. "Report of the WHO-China joint mission on coronavirus disease 2019 (COVID-19).

www.who.int/docs/default-source/coronaviruse/who-china-joint-mission-on-covid-19-final-report.pdf

13. WHO. "Novel coronavirus (2019-nCoV) situation report – 1."

www.who.int/docs/default-source/coronaviruse/situation-reports/20200121-sitrep-1-2019-ncov.pdf?sfvrsn=20a99c10_4

CHAPTER FOUR

1. The Independent Panel. "Make it the Last Pandemic."

https://theindependentpanel.org/wp-content/uploads/2021/05/COVID-19-Make-it-the-Last-Pandemic_final.pdf

2. WHO. "Novel coronavirus – Republic of Korea (ex-China)."

www.who.int/csr/don/21-january-2020-novel-coronavirus-republic-of-korea-ex-china/en/

3. Reuters. "Special Report: How Korea trounced U.S. in race to test people for coronavirus."

www.reuters.com/article/us-health-coronavirus-testing-specialrep/special-report-how-korea-trounced-u-s-in-race-to-test-people-for-coronavirus-idUSKBN2153BW

4. SCMP. "Coronavirus cluster emerges at another South Korean church, as others press ahead with Sunday services."

www.scmp.com/week-asia/health-environment/article/3077497/coronavirus-cluster-emerges-another-south-korean

5. WHO. "Coronavirus disease 2019 (COVID-19) situation report – 86."

www.who.int/docs/default-source/coronaviruse/situation-reports/20200415-sitrep-86-covid-19.pdf?sfvrsn=c615ea20_6

6. WHO. "Coronavirus disease (COVID-19) situation report – 116."

   www.who.int/docs/default-source/coronaviruse/situation-reports/20200515-covid-19-sitrep-116.pdf?sfvrsn=8dd60956_2

7. SCMP. "Coronavirus : Infected Chinese tourist in France dies, in Europe's first death."

   www.scmp.com/news/world/europe/article/3050802/coronavirus-infected-chinese-tourist-france-dies-europes-first

8. The Guardian. "Coronavirus:British nine-year-old in hospital in France."

   www.theguardian.com/world/2020/feb/08/coronavirus-five-new-cases-in-france-are-british-nationals

9. Wikipedia. "Coronavirus pandemic in France"

   https://en.wikipedia.org/wiki/2020_coronavirus_pandemic_in_France

10. WHO. "Coronavirus disease (COVID-19) situation report –86.

    https://www.who.int/docs/default-source/coronaviruse/situation-reports/20200415-sitrep-86-covid-19.pdf

11. WHO. "Coronavirus disease (COVID-19) situation report – 116."

    www.who.int/docs/default-source/coronaviruse/situation-reports/20200515-covid-19-sitrep-116.pdf?sfvrsn=8dd60956_2

12. Reuters. "Italy's coronavirus outbreak infects 51 people, kills two."

www.reuters.com/article/us-china-health-italy/second-coronavirus-patient-dies-in-italy-ansa-news-agency-idUSKCN20G0CQ

13. New York Times. "Boris Johnson contracts coronavirus, rattling top ranks of U.K. Government."

    www.nytimes.com/2020/03/27/world/europe/boris-johnson-coronavirus.html

14. The Local. "Update: Five new coronavirus cases confirmed in western Germany." Published February 26, 2020.

    www.thelocal.de/20200226/two-new-coronavirus-cases-confirmed-in-germany

15. The Chicago Tribune. "Chicago woman who travelled to China diagnosed with coronavirus, health officials say."

    www.chicagotribune.com/news/breaking/ct-coronavirus-china-epidemic-illinois-case-20200124-yx2xd3yeovar3o25ei6bfvvbze-story.html

16. Live Science. "First US death from COVID-19 reported in Washington state."

    www.livescience.com/coronavirus-spread-washington-nursing-home.html

17. Business Insider. "One chart shows New York City's coronavirus cases, deaths, and hospitalizations by age bracket as the city's cases top 20,000"

    www.businessinsider.in/science/news/one-chart-shows-new-york-citys-coronavirus-cases-deaths-and-hospitalizations-by-age-bracket-as-the-citys-cases-top-20000/articleshow/74837730.cms

18. "First novel coronavirus case reported in Australia."

    https://tass.com/world/1112907

CHAPTER FIVE

1. Li Q, Guan X, Wu P, et al., "Early transmission dynamics in Wuhan, China, of novel coronavirus–Infected pneumonia." *New England Journal of Medicine*. doi:10.1056/NEJMoa2001316
2. WHO. "Coronavirus disease 2019 (COVID-19) situation report–48."

   www.who.int/docs/default-source/coronaviruse/situation-reports/20200308-sitrep-48-covid-19.pdf?sfvrsn=16f7ccef_4
3. Cascella M, Rajnik M, Cuomo A, et al., "Features, Evaluation and Treatment of Coronavirus (COVID-19)"

   www.ncbi.nlm.nih.gov/books/NBK554776/
4. Huang C, Wang Y, Li X, et al. "Clinical features of patients infected with 2019 novel coronavirus in Wuhan, China." *Lancet* Published February 15, 2020; 395(10223): 497–506
5. WHO. "Prevention and control during health care when novel coronavirus (nCoV) infection is suspected."

   www.who.int/publications-detail/infection-prevention-and-control-during-health-care-when-novel-coronavirus-(ncov)-infection-is-suspected-20200125
6. FDA. "(COVID-19) update: FDA issues emergency use authorization for potential COVID-19 treatment."

   www.fda.gov/news-events/press-announcements/coronavirus-covid-19-update-fda-issues-emergency-use-authorization-potential-covid-19-treatment
7. WHO. "Home care for patients with COVID-19 presenting with mild symptoms and management of their contacts."

   www.who.int/publications-detail/home-care-for-patients-with-suspected-novel-coronavirus-(ncov)-infection-presenting-with-mild-symptoms-and-management-of-contacts

8. WHO. "Infection prevention and control during health care when novel coronavirus (nCoV) infection is suspected." Published March 19, 2020.

   https://www.who.int/publications-detail/infection-prevention-and-control-during-health-care-when-novel-coronavirus-(ncov)-infection-is-suspected-20200125

9. WHO "Infection prevention and control of epidemic- and pandemic-prone acute respiratory diseases in health care." Published March 29, 2020

   https://www.who.int/publications/i/item/infection-prevention-and-control-of-epidemic-and-pandemic-prone-acute-respiratory-infections-in-health-care

10. WHO. 2008 "How to put on and take off Personal Protective Equipment (PPE)."

    www.who.int/csr/resources/publications/putontakeoffPPE/en/

CHAPTER SIX

1. The Guardian. "136m Americans now unemployed as another 3m file for benefits."

   www.theguardian.com/business/2020/may/14/unemployment-us-data-coronavirus

2. WHO. "Physical and mental health key to resilience during COVID-19 pandemic."

   www.euro.who.int/en/media-centre/sections/statements/2020/statement-physical-and-mental-health-key-to-resilience-during-covid-19-pandemic

3. The Hindu. "Coronavirus: Hong Kong Disneyland to be

closed to help prevent spread of virus."

www.thehindu.com/news/international/coronavirus-hong-kong-disneyland-to-be-closed-to-help-prevent-spread-of-virus/article30656912

4. London Book Fair. "Book Fair cancelled"

   www.londonbookfair.co.uk/

5. The Verge. "Geneva Motor Show cancelled after coronavirus causes government to ban large events."

   www.theverge.com/2020/2/28/21156368/geneva-motor-show-2020-canceled-coronavirus-concerns

6. Olympic. "Joint Statement from the International Olympic Committee and the Tokyo 2020 Organising Committee."

   www.olympic.org/news/joint-statement-from-the-international-olympic-committee-and-the-tokyo-2020-organising-committee

7. Statista. "Forecasted change in revenue from the travel and tourism industry due to the coronavirus (COVID-19) pandemic worldwide from 2019 to 2020."

   www.statista.com/forecasts/1103426/covid-19-revenue-travel-tourism-industry-forecast

8. National Herald. "Coronavirus may knock 1.3 per cent off global GDP growth."

   www.nationalheraldindia.com/opinion/coronavirus-may-knock-13-per-cent-off-global-gdp-growth

9. OECD. "OECD Interim Economic Assessment – Coronavirus: The world economy at risk."

   www.oecd.org/berlin/publikationen/Interim-Economic-

Assessment-2-March-2020.pdf

10. Vinelli A, Weller CE, Vijay D, "The economic impact of coronavirus in the U.S. and possible economic policy responses."

    www.americanprogress.org/issues/economy/news/2020/03/06/481394/economic-impact-coronavirus-united-states-possible-economic-policy-responses/

11. World Economic Forum. "This is how much the coronavirus will cost the world's economy, according to the UN."

    www.weforum.org/agenda/2020/03/coronavirus-covid-19-cost-economy-2020-un-trade-economics-pandemic

12. "Economic Outlook, April 2020: The great lockdown."

    www.imf.org/en/Publications/WEO/Issues/2020/04/14/weo-april-2020

CHAPTER SEVEN

1. WHO. "Coronavirus disease (COVID-19) advice for the public."

   www.who.int/emergencies/diseases/novel-coronavirus-2019/advice-for-public

2. CDC. "How to protect yourself."

   www.cdc.gov/coronavirus/2019-ncov/prevent-getting-sick/prevention.html

3. Economic Times. "How long can the coronavirus survive on surfaces or in the air?"

   https://economictimes.indiatimes.com/magazines/panache/how-long-can-coronavirus-live-on-surfaces-or-in-the-air/articleshow/74690737.cms?from=mdr

4. Beall A, "Why social distancing might last for some time" *BBC*. Published March 25, 2020.

   www.bbc.com/future/article/20200324-covid-19-how-social-distancing-can-beat-coronavirus

5. Cetron M, Landwirth J. 2005 "Public health and ethical considerations in planning for quarantine." *Yale Journal of Biology and Medicine* 78(5): 329–334

6. "Ethical and legal considerations in mitigating pandemic disease: Workshop summary"

   www.ncbi.nlm.nih.gov/books/NBK54163/

7. WHO. "International health regulations and epidemic control."

   www.who.int/trade/distance_learning/gpgh/gpgh8/en/index7.html

CHAPTER EIGHT

1. Jennifer Robinette. "The Science Behind COVID-19 Vaccines." *St. Jude Children Research Hospital*

   https://www.stjude.org/about-st-jude/stories/making-a-difference/science-behind-covid-19-vaccines.html

2. Florian Krammer. "SARS-CoV-2 Vaccines in Development." Published September 23, 2020. *Nature*

   https://www.nature.com/articles/s41586-020-2798-3

3. CDC. "COVID-19 Treatments and Medications." Updated August 5, 2022.

   https://www.cdc.gov/coronavirus/2019-ncov/hcp/therapeutic-options.html

4. CDC. "Understanding mRNA COVID-19 Vaccine." Updated July 15, 2022.

https://www.cdc.gov/coronavirus/2019-ncov/vaccines/different-vaccines/mrna.html

5. M. E. Halloran, I. M. Longini Jr., C. J. Struchiner, "Design and Analysis of Vaccine Studies." *Springer*, 2010.

CHAPTER NINE

1. Hopkins Medicine. "COVID-19 Variants: What you should know." Updated April 8, 2022.

https://www.hopkinsmedicine.org/health/conditions-and-diseases/coronavirus/a-new-strain-of-coronavirus-what-you-should-know

2. WHO. "Tracking SARS-CoV-2 Variants". Updated August 11, 2022.

https://www.who.int/activities/tracking-SARS-CoV-2-variants

3. Mayo Clinic. "COVID-19 Variants: What's the Concern?" Published August 27, 2022.

https://www.mayoclinic.org/diseases-conditions/coronavirus/expert-answers/covid-variant/faq-20505779

4. Mayo Clinic. "HOW Emerging Variants Could Affect COVID-19 Variants and Spread." Published February 12, 2021.

https://newsnetwork.mayoclinic.org/discussion/how-emerging-variants-could-affect-covid-19-testing-vaccines-and-spread/

5. Mayo Clinic. "What Do I Need to Know About the Delta Variant?" Published September 4, 2021.

https://www.mayoclinic.org/what-do-i-need-to-know-about-the-delta-variant/vid-20519610

CHAPTER TEN

1. Sarah Zang; DEC 8, 2021, The Atlantic

2. Joe Pinker; DEC 6, 2021, Atlantic Daily

3. NADIA A. SAM-AGUDU, BOGHUMA KABISEN TITANJI, FREDROS OKUMU, AND MADHUKAR PAI MAR 4, 2022, The Atlantic

4. HOPE: By Jared S. Hopkins and Joseph Walker Wall Street Journal.

5. "US No Longer in 'Full-blown' Pandemic Phase, Fauci says."

https://www.washingtonpost.com/health/2022/04/27/pandemic-phase-over-fauci-covid/

6. New York Post. " Dr. Anthony Fauci walks back comment about US no longer being in 'pandemic phase'." Published August 29, 2022

https://nypost.com/2022/04/27/dr-fauci-walks-back-comment-about-us-no-longer-in-pandemic-phase/

7. Ed Yong. "Pulitzer Prize for Explanatory Reporting for coverage of the COVID-19 pandemic." *The Atlantic*

EPILOGUE

1. Le TT, Andreadakis Z, Kumar A, et al., "The COVID-19 vaccine development landscape." *Nature*. Published April 9, 2020.

www.nature.com/articles/d41573-020-00073-5

2. CSIS. "Ending the cycle of crisis and complacency in U.S. Global Health Security." Published November 18, 2019.
https://healthsecurity.csis.org/final-report/

3. "WHO continues to track the evolving infectious disease situation, sound the alarm when needed, share expertise, and mount the kind of response needed to protect populations from the consequences of epidemics, whatever and wherever might be their origin."
www.who.int/csr/alertresponse/en/

4. CDC Foundation. "What is public health? CDC Foundation."
www.cdcfoundation.org/what-public-health

5. WHO. "COVID-19 reveals gaps in health systems: WHO briefing."
www.weforum.org/agenda/2020/05/covid-19-reveals-gaps-in-public-health-system-who-briefing/

6. The Independent. "Government 'systematic underfunding' to blame for NHS humanitarian crisis, Labour says."
www.independent.co.uk/news/uk/politics/nhs-humanitarian-crisis-red-cross-labour-jeremy-hunt-underfunding-spending-cuts-a7514626.html

7. Lancet. "Gap in India's preparedness for COVID-19 control, Patralekha Chatterjee." www.thelancet.com/journals/laninf/article/PIIS1473-3099(20)30300-5/fulltext

8. Times of India. "Dry S, Have to learn to live with COVID-19: Govt" https://timesofindia.indiatimes.com/india/have-to-learn-to-live-with-covid-19-govt/articleshow/75638429.cms

9. Eliza Mackintosh. "The 'Worst Variant' is Here" *CNN*. Last updated July 14, 2022 https://apple.news/ATHc4E233Sz-Ri_aqQOVkSw

Made in United States
Orlando, FL
04 November 2026

90201004R00115